D1615218

# Geomorphology in Environmental Planning

British Geomorphological Research Group
Symposia Series

*Geomorphology in Environmental Planning*

Edited by
**J. M. Hooke**

# Geomorphology in Environmental Planning

*Edited by*
**J. M. Hooke**
*Department of Geography*
*Portsmouth Polytechnic*

John Wiley and Sons
CHICHESTER · NEW YORK · BRISBANE · TORONTO · SINGAPORE

***Library of Congress Cataloging in Publication Data:***
Geomorphology in environmental planning.
(Symposia series / British Geomorphological Research Group)
Papers presented at a symposium held at the Institute of British Geographers' Conference at Portsmouth in Jan. 1987.
1. Geomorphology—Congresses. I. Hooke, J. M. (Janet M.) II. Institute of British Geographers. Conference (1987 : Portsmouth, Hampshire) III. Series: Symposia series (British Geomorphological Research Group).
GB400.2.G444 1988 551.4 88-5557
ISBN 0 471 91988 8

***British Library Cataloguing in Publication Data:***
Geomorphology in environmental planning.
1. Geomorphology—For environmental planning
I. Hooke, J. M. (Janet M.) II. Series
551.4′02471

ISBN 0 471 91988 8

Phototypeset by Dobbie Typesetting Service, Plymouth, Devon
Printed by The Bath Press, Bath, Avon

To my father

# Contents

**River Management**

# Series Preface

The British Geomorphological Research Group (BGRG) is a national multidisciplinary Society whose object is 'the advancement of research and education in Geomorphology'. Today, the BGRG enjoys an international reputation and has a strong membership from both Britain and overseas. Indeed, the Group has been actively involved in stimulating the development of Geomorphology and Geomorphological societies in several countries. The BGRG was constituted in 1961 but its beginnings lie in a meeting held in Sheffield under the chairmanship of Professor D. L. Linton in 1958. Throughout its development the Group has sustained important links with both the Institute of British Geographers and the Geological Society of London.

Over the past three decades the BGRG has been highly successful and productive. This is reflected not least by BGRG Publications. Following its launch in 1976 the Group's Journal, *Earth Surface Processes* (since 1981 *Earth Surface Processes and Landforms*) has become acclaimed internationally as a leader in its field, and to a large extent the Journal has been responsible for advancing the reputation of the BGRG. In addition to an impressive list of other publications on technical and educational issues, BGRG symposia have led to the production of a number of important works including *Nearshore sediment dynamic and sedimentation* edited by J. R. Hails and A. P. Carr; *Geomorphology and Climate* edited by E. Derbyshire; *River Channel Changes* edited by K. J. Gregory; and *Timescales in Geomorphology* edited by R. Cullingford, D. Davidson and J. Lewin. This sequence of books culminated 1987 with a publication of the *Proceedings of the First International Geomorphology Conference* edited by Vince Gardiner. This international meeting, arguably the most important in the history of Geomorphology, provided the foundation for the development of Geomorphology into the next century.

This open-ended BGRG Symposia Series has been founded and is now being fostered to help maintain the research momentum generated during the past three decades, as well as to further the widening of knowledge in component fields of geomorphological endeavour. The Series consists of authoritative volumes based on the themes of BGRG meetings, incorporating, where appropriate, invited contributions to complement chapters selected from presentations at these meetings under the guidance and editorship of one or more suitable specialists. Whilst maintaining a strong emphasis on pure geomorphological research. BGRG meetings are diversifying, in a very positive way, to consider links between geomorphology *per se* and other disciplines such as ecology, agriculture, engineering and planning. This first volume in the Series on *Geomorphology in Environmental Planning*, edited by Janet Hooke, exemplifies this new trend.

The series will contribute to advancing geomorphological research and we look forward to the effective participation of geomorphologists and other scientists concerned with earth surface processes and landforms, their relation to Man, and their interaction with the other components of the Biosphere.

Geoffrey Petts
BGRG Publications

28 March 1988

# Preface

The interface between geomorphology and environmental planning is potentially a very exciting one. Geomorphologists can provide techniques of investigation and knowledge of the earth's surface, its forms and processes. Such information is vital to planning of the environment to prevent development in unsuitable areas, to reduce damaging consequences of activities and to predict the effects of policies.

Effective interaction requires an awareness on the part of planners and policy-makers of the expertise offered by geomorphologists and an awareness by geomorphologists of the policy frameworks, decision-makers and processes of policy-formulation. Details on each of these aspects are provided in this book and the advantages of this mutual awareness are demonstrated clearly. Strategies for involvement are suggested though the difficulties are also indicated. Policy voids and data needs are identified.

The papers included here were amongst those presented at a symposium held at the Institute of British Geographers' Conference at Portsmouth in January, 1987. The volume follows a similar format; in each major geomorphological or policy sphere a review is provided by a leading authority and this is followed by a case study of a specific application or problem.

The book is aimed at geomorphologists who wish to see their research results used more fruitfully and who wish to have a professional influence on environmental policies, and at environmental planners and policy-makers to demonstrate the value and necessity of using geomorphology to achieve sound environmental management. It encompasses issues in rural and urban areas and dynamic zones such as rivers and coasts. The papers focus on policy in the United Kingdom but draw on experience elsewhere.

I am very grateful to all who have helped in production of this volume but above all to Rob for his unerring support. In the end the gestation was longer than that of our daughter but both arrived eventually.

Janet Hooke

*Portsmouth*

# List of Contributors

**Richard Aspinall**, The Macaulay Land Use Research Institute, Craigiebuckler, Aberdeen AB9 2QJ.

**Dr John Boardman**, Countryside Research Unit, Department of Humanities, Brighton Polytechnic, Falmer, Brighton, East Sussex BN1 9PH.

**Dr David Brook**, Land Stability Branch, Minerals Division, Department of the Environment, 2 Marsham Street, London SW1P 3EB.

**Dr Andrew Brookes**, Thames Water, Nugent House, Vastern Road, Reading RG1 8DB

**Professor Denys Brunsden**, Department of Geography, King's College London, The Strand, London WC2R 2LS.

**Dr Alan Carr**, 17 Pyles Thorne Close, Wellington, Somerset TA21 8EF.

**Dr Bob Crabtree**, WRc Engineering, Frankland Road, Blagrove, PO Box 85, Swindon, Wiltshire SN5 8YR.

**Professor Ian Douglas**, School of Geography, University of Manchester, Manchester M13 9PL.

**Peter Gagen**, Department of Environmental and Geographical Studies, Manchester Polytechnic, John Dalton Building, Chester Street, Manchester M1 5GD.

**Professor Ken Gregory**, Department of Geography, University of Southampton, Southampton SO9 5NH.

**Dr John Gunn**, Department of Environmental and Geographical Studies, Manchester Polytechnic, John Dalton Building, Chester Street, Manchester M1 5GD

**Dr Janet Hooke**, Department of Geography, Portsmouth Polytechnic, Buckingham Building, Lion Terrace, Portsmouth PO1 3HE.

**Dr Mark Macklin**, Department of Geography, University of Newcastle upon Tyne, Newcastle upon Tyne NE1 7RU.

**Dr Brian Marker**, Land Stability Branch, Minerals Division, Department of the Environment, 2 Marsham Street, London SW1P 3EB.

**Professor Roy Morgan**, Silsoe College, Cranfield, Silsoe, Bedford MK45 4DT.

**Professor Malcolm Newson**, Department of Geography, University of Newcastle upon Tyne, Newcastle upon Tyne NE1 7RU.

**Dr Stan Openshaw**, Department of Geography, University of Newcastle upon Tyne, Newcastle upon Tyne NE1 7RU.

**Dr Dennis Parker**, Flood Hazard Research Centre, Middlesex Polytechnic, Queensway, Enfield, Middlesex EN3 4SF.

**Professor Edmund Penning-Rowsell**, Flood Hazard Research Centre, Middlesex Polytechnic, Queensway, Enfield, Middlesex EN3 4SF.

**Dr Jane Rickson**, Silsoe College, Cranfield, Silsoe, Bedford MK45 4DT.

**Dr Peter Sims**, Department of Geographical Sciences, Plymouth Polytechnic, Drake Circus, Plymouth, Devon PL4 8AA.

**Dr Les Ternan**, Department of Geographical Sciences, Plymouth Polytechnic, Drake Circus, Plymouth, Devon PL4 8AA.

**Paul Thompson**, Flood Hazard Research Centre, Middlesex Polytechnic, Queensway, Enfield, Middlesex EN3 4SF.

# Introduction

Geomorphology in Environmental Planning
Edited by J. M. Hooke



# 1 Introduction: Frameworks for Interaction

**JANET HOOKE**
*Department of Geography, Portsmouth Polytechnic*

## INTRODUCTION

This book examines the interaction between geomorphology and public policies. The concern is with policy areas which affect the physical landscape and the processes acting within it. It focuses not only on the impact of present policies but also on identification of policy needs and on the design and implementation of policies. The nature of environmental problems in affecting everyone means that most such policies are discussed, formulated and implemented largely within the public domain, though the response of individuals may have a profound effect on the resulting situation.

Public policies are implemented through a range of measures varying from strict legislation and regulations to vague guidelines and financial incentives. The public policies may apply at different levels from international to local though most of the basic frameworks are laid down at national level in accordance with the policies of the Government in power. Some fulfilment of the policies is overseen by government ministries but many of the policies are actually carried out by other government agencies and by local authorities. In Britain the policies of the European Community have an increasing influence.

Geomorphologists are concerned with the form and processes of the earth's surface so any activity which modifies the shape of the land, induces movement of materials or alters the quantity or quality of water and drainage, is of interest to them. Many activities may indirectly affect the earth surface properties through interaction with vegetation. Coates (1984) in discussing geomorphology and public policy says:

> Since humans live, work, build and play on the earth's surface such endeavours inevitably produce changes to the land–water ecosystem. Thus it is common for geomorphologists to take a special interest in those societal activities that are designed to modify natural processes. . . . Therefore it is altogether appropriate and necessary that geomorphologists become involved with those in government who are authorized to make

decisions which may affect the character of the topography and the materials and processes that comprise it.

Environmental changes due to human activities have taken place in the past but the rate of change and the ability of humans to alter the landscape is increasing. Also, as population increases and resources diminish the pressure on the remaining resources grows. Increasingly it is realised that there may be detrimental consequences in the environment if uncontrolled development takes place and the rate continues to accelerate. The combination of these factors together with a greater knowledge and understanding of the environment has meant an increased awareness of and concern with environmental problems. This is witnessed by the growth in numbers and membership of environmental pressure groups and environmental interest groups and by the numbers and sales of environmental publications.

The attitudes to the environment still vary considerably and are much discussed in the literature. Arguments still tend to be economically based but the need to incorporate environmental costs in evaluation of projects is realised increasingly and the subject of environmental economics is expanding. Legislation requiring Environmental Impact Assessments of projects is already in operation in the United States and guidelines are being introduced in the European Community.

The purpose of this book is to examine the geomorphological impacts of present policies and discuss the implications for future policies and policy modifications. It aims also to examine how policy formulation takes place, to demonstrate how geomorphologists can become involved in the process and to identify the areas where policy needs to be formulated. For policy-makers and landscape managers it demonstrates the value and necessity of geomorphological information in examining environmental problems and designing management strategies. The focus in this book is on Britain; however, some examples are taken from other countries whose experience may help in understanding the effects of policies and in evaluating the alternatives.

## TYPES OF CHANGES AND PROBLEMS

Some geomorphological changes are the result of deliberate and large-scale alteration to the earth's surface as in the case of land excavation or quarrying and mining. In other cases the most obvious results of an activity or policy may be aesthetic and ecological and the geomorphological changes are more subtle. Issues of public policy having landscape impacts can be considered in terms of major types of human activity, geomorphological zones or policy spheres.

Probably some of the most obvious changes taking place in the British landscape at present and some of the most controversial in terms of policies are those in rural areas related to land use. The change in type of agriculture with greater emphasis, particularly in the lowlands, on increased cereal production, and the incentives for greater use of the land which have led to removal of hedges, drainage of areas and ploughing of moorlands, have all had profound effects on the appearance of the landscape and also on the amounts of soil erosion and the quantity and quality of water. The effects are manifest not only in the source areas but also downslope and downstream. Similarly, present policies of increasing forestry are controversial and the consequences of such policies need to be examined.

Although alteration of the landscape is perhaps more obvious and also less readily accepted in rural areas, the changes in the urban areas are, at least as great. These are the areas where most people live though the environment is in many ways more tightly controlled. Often the geomorphological consequences of urban development are ignored because of the assumption of control in the man-made environment and because of lack of knowledge of the natural systems and processes still operating in such areas. Suitable policies of control are indeed often lacking and public authorities are forced to pick up the consequences of private actions such as building development.

Certain activities may be either rural or urban and involve direct morphological alteration of the land. These include road and building construction and the mining and quarrying of materials. Both tend to be destructive of land but complementary activities of solid waste disposal and land reclamation may also have a significant impact on the landscape. The latter topics appear relatively neglected by geomorphologists. Waste disposal in the form of river pollution and also the interactions of acid rain have received some attention.

There are two spatial zones, rivers and coasts, which are especially dynamic geomorphologically and therefore present particular problems. Problems in these zones are also caused by lack of coordinated policies and by the proliferation of agencies involved. It is not only morphological changes which may be significant but also alterations of processes and water quality.

Geomorphological problems and policies could be examined in terms of particular types of area, as recognized by legislation, which are subject to special policy and planning treatment in the landscape. Such areas include, for example, National Parks but those are not singled out here; to a large extent the problems reflect in a more concentrated way the controversies in other parts of the landscape. Another major area of geomorphological work is that of natural hazards and interaction with policy but this is not discussed explicitly within this volume. This is partly because, compared with environments elsewhere in the world, Britain is not particularly hazardous geomorphologically. The major hazards are from river and coastal flooding and landslides and these are implicated in the other discussions. Many of the hazards now encountered in Britain are anthropogenic.

## EXISTING FRAMEWORKS

Policy can be enacted at a variety of levels and in a number of different ways, exerting different degrees of control varying from binding legislation to vague guidelines and incentives for certain practices to be adopted. Park (1986) identifies three policy mechanisms—moral persuasion, direct (legal) regulation, and economic incentives. The national context generally provides the overall framework in which the general aims of the policy are identified and the instruments of implementation articulated. This level may now be influenced by international pressure as is seen in the case of air and marine pollution, and even by legislation, particularly from the European Community in the case of Britain. As pressure on resources becomes greater, so the need for legislation appears to be increasing. Environmental policies have now become election issues.

### Legislation

Present environmental legislation is covered by a number of different Acts in Britain but the main ones of concern encompass planning in both urban and rural areas, agriculture and forestry, special areas such as National Parks, mineral development, rivers and coasts, and wildlife and countryside. In rural areas there appear to be three main areas of legislation—urban development pressures, resources and conservation (Gilg, 1986; Open University, 1985) and these are now outlined. It is impossible here to give details of all relevant legislation; information on frameworks relating to individual policy areas are given in the chapters that follow.

The Town and Country Planning Act provided the basic framework of controls in 1947 by requiring planning permission to be granted for various developments to take place. It tried to identify non-conflicting land use zones. The 1971 Act provided the present basis for both creation of plans and development control. This required Structure Plans to be prepared by County Councils and Local Plans to be drawn up by District Councils. Development control implemented by planning permission is exercised at local level. Public Inquiries can be held into issues of national or regional importance or outside the scope of local planning authorities (Gilg, 1986). Three special areas in which development should be controlled have also been identified; these are green belts, National Parks and Areas of Outstanding Natural Beauty. In these areas planning permission is harder to obtain. However, in the latter two legislation does not cover agriculture and forestry and in recent years pressure for development of these activities, particularly in the uplands, has led to severe conflict, for example in ploughing of moorlands on Exmoor. Mineral working is covered by planning legislation.

In terms of resources, particularly of the countryside, the main areas of legislation and controls are in agriculture, forestry, rivers and coasts.

Agricultural policy has been aimed at maximizing production of food in Britain and minimizing the need for imports. The Common Agricultural Policy (CAP) of the European Community now provides the framework for farming practices. There are two areas of operation influencing farming; (a) guarantee section in which prices are assured to the farmer and excess production is taken into store, (b) guidance section in which grants and incentives are provided for particular activities and in particular areas. It is under this section that the Less Favoured Areas are recognized which as Newson shows (Chapter 2) has affected the British uplands. The philosophy underlying forestry policies has also been one of expansion. The policies of expansion of agriculture and forestry are now being questioned partly because of the cost and morality of unused agricultural surpluses and partly because of the effects on the landscape.

Under the 1973 Water Act the Water Authorities were created in England and Wales and were given responsibilities for water supply, sewerage and sewage disposal, recreation and some conservation. They also have a responsibility for sea defences to prevent coastal flooding. Legislative and institutional frameworks for coastal management are rather more disparate as indicated in the papers by Carr (Chapter 11) and Penning-Rowsell, Thompson and Parker (Chapter 12) and responsibility for coast protection has recently transferred from the Department of the Environment (DOE) to the Ministry of Agriculture, Fisheries and Food (MAFF).

In terms of conservation the relevant legislation at present is the Wildlife and Countryside Act of 1981. Since agriculture is not subject to planning controls unlike other developments, the aim was to reduce conflict between agricultural and conservation interests by providing some protection for SSSIs (Sites of Special Scientific Interest) and to allow for management agreements to be entered into with landowners. One of the major problems has been lack of enough money to pay compensation for development foregone, but in several areas farmers have agreed to retain or revert to grazing land and to use more traditional agricultural techniques.

## Organizations and Agencies

Policy-formulation, national decision-making and legislation in Britain is in the hands of the Government and Parliament. Gilg (1986) outlines the hierarchy of government and Park (1986) indicates a hierarchy of scales of concern from local through national and transfrontier to international. Executive and advisory agencies may both give advice and be responsible for implementation and management. Policy formulation is now also influenced by pressure groups.

There are a number of bodies which are international in scope and organization some of which have influenced environmental policies and have certainly been responsible for implementation and coordination of policies. These include the United Nations and its various agencies and the World Bank.

At the national level Governmental Ministries are ultimately responsible for environmental policies, of which a major one is obviously the Department of the Environment, but the Ministry of Agriculture, Fisheries and Food (MAFF) also has some control over water resources and coastal management as well as agriculture. In several cases it is found that separate ministries and bodies have conflicting policies illustrating the lack of coordination in the environmental field.

There are several other bodies which have a major role in formulating and/or implementing environmental policies. Many of these bodies are quangos with appointed rather than elected members. Most are staffed with professionals from environmental subjects. These include the Nature Conservancy Council whose major functions are the establishment and maintenance of National Nature Reserves and SSSIs, advice to the government on nature conservation policy, and provision of information and advice to interested parties. Their responsibilities include geological and geomorphological aspects of nature conservation as well as ecological aspects. The Countryside Commission, another quango, has three main responsibilities: conserving the landscape beauty of the countryside, developing and improving facilities for recreation and access in the countryside, and advising government on matters of countryside interest in England and Wales. Other bodies concerned with particular sectors include the Forestry Commission which is responsible for research and advice about forestry as well as owning and managing areas of forest. The Water Authorities of England and Wales are responsible for the management of water services, water distribution, pollution, sewerage, sea defences, recreation and inland fisheries. Through the Water Authorities Association they advise Ministers on policies in these areas. There are obviously many other bodies whose actions may have an impact on the landscape and environment. Further information on the agencies and their functions is found in the *Countryside Handbook* (Open University, 1985).

Local planning functions, and some regulation and management of the environment is in the hands of the local authorities, of which in Britain there are two main tiers, County and District Councils. They have some scope for policy decision-making and can also influence central Government to some extent through the Associations of County and District Councils.

The actual implementation of policies, their impact, effects and efficacy have long depended to a large extent on individual responses, particularly where the policy is one of incentives or persuasion rather than regulation. This is perhaps most obvious in the case of farming. As Blunden and Curry (1985) state:

> The use which agriculture makes of the countryside . . . is not, in most cases, the result of deliberate and coordinated planning. Rather it is the result of a myriad set of business decisions taken by individual farmers. In turn, they are responding within the constraints imposed by the soil

types, climate, location, other technical limitations and personal preferences and objectives to signals coming to them predominantly in the form of prices of farm products and costs of inputs.

Increasingly, private and pressure groups have had an influence on environmental policies. Their influence does vary considerably depending on the group and also the political party in power but for example, the National Farmers Union and the Country Landowners Association have proved very powerful in the agricultural and conservation field (or anti-conservation). Many groups have grown in membership and influence over the past decade or so and important groups include the National Trust, the Civic Trust, Royal Society for the Protection of Birds, Royal Society for Nature Conservation, Friends of the Earth, Council for the Preservation of Rural England, British Trust for Conservation Volunteers and Greenpeace. The latter is international and there are some other private bodies at international level who may influence policies including multinational corporations and aid organizations. Some of these may be in favour of development and against environmental considerations.

In many aspects of environmental concern policies may not have been formulated and management is carried out through accepted and agreed practices. Many ancient rights are of this nature and scarce resources were often allocated in this way originally. Many guidelines and policies are also open to wide interpretation and so local practices develop. In the field of water management for example, this can be apparent in practices of clearing rivers of bars and trees, a practice which is difficult to justify on hydrological or economic grounds but which is long established in many areas. Although such work may be carried out by public bodies there is often a lack of accountability of practices at this level.

## POLICY FORMULATION

Much policy in the environmental field is crisis-led. No policy exists until there has been some disaster when legislation is often hastily enacted. As Gilg (1986) has said, 'Environmental and management policies in Britain have often changed only under the pressure of events and changed circumstances'. This situation may be altering as environmental awareness grows; as the impacts of activities in the environment are demonstrated, there is a growing realization that preventative action may be needed. The steps and pathways in policy formulation depend on the field and the organization involved but Coates (1984) has identified four components of decision-making:

(1) Perception phase—a perceived need for action.
(2) Planning phase.
(3) Implementation phase.
(4) Management phase.

If geomorphologists are to participate in policy-formulation then they must help to demonstrate the need for a policy and then suggest effective action for the later stages. According to Coates (1984) to proceed to policy adoption the specific objectives must be identified, alternative solutions must be evaluated and then the option with the greatest chance of success must be selected.

Several authors have pointed out in their chapters that it is insufficient to provide data in a vacuum but rather advice and information must be orientated to the particular problem, must be assessed within the socio-economic and political frameworks and must be directed to the right people. It is therefore important to know what the frameworks are and the nature of the links and influences in a system. On this Coates (1984) says 'it is insufficient to simply write a publication in the hope that policy makers will read it and translate its findings into appropriate action. . . . The lack of communication between scientists and policy makers can negate careful investigation and research'. O'Riordan (1981) speaks of the 'charade of objectivity'—rarely is work entirely independent or without subjective premises, it is inevitably set within some framework. The framework often precludes certain lines of investigation and answers and leads towards certain types of result.

It is interesting perhaps, as a preface to the other chapters, to examine the environmental policy issues recognized by non-geomorphologists. These can be largely grouped into the rural land resource base, urban development pressures and reconciling development and conservation (Gilg, 1986). Within these O'Riordan (1981) has identified countryside conservation in relation to rural land resources, nuclear power and pollution in urban pressures and Environmental Impact Assessment and UK conservation strategy as particular problems. Blunden and Curry (1985), in discussing issues in the countryside, reproduce a table from the Countryside Review Committee in 1976 in which impacts and conflicts of existing policies are identified. The Countryside Commission has recently published a major review of the changing landscape (Countryside Commission, 1987) and their Countryside Policy Review Panel has examined issues related to rural land use, agriculture, forestry, recreation and the urban fringe, particularly examining the integration of farming and environmental conservation. The Economic and Social Research Council (ESRC) Environment and Planning Committee in identifying research subjects 'likely to be of enduring public concern, most in need of a multidisciplinary input from the social sciences' (and that would also provide the most effective stimulus to the development of research methodology) selected rural land use, risk, environmental economics, and energy and the environment (ESRC, 1987). Table 1.1 lists the seven major projects established.

Briggs (1986) identifies three main threads in European Community policies—reducing pollution, conserving natural resources, and protecting the environment. Pressure groups vary in their particular interest in and concern for the environment but impact of agricultural policies, draining of wetlands

**Table 1.1** Major research projects initiated by ESRC on environmental issues

1. Responsiveness of cereal producers to changing policy and incentives.
2. Set aside as an environmental and agricultural policy instrument.
3. Occupancy change and the farmed landscape.
4. Fiscal measures and farm management.
5. Hazardous waste regulations in the UK and EC—comparison of risk management
6. Integrating the environment with energy supply and demand decisions.
7. Energy conservation through interventionist measures.

and problems of air and water quality and waste disposal have figured very large in recent years. At the time of writing there is much public debate about policy on future use of agricultural land now that it is realized that production may have to be decreased and incentives given for alternative uses of the land. Much of the debate centres around the extent to which urban development, leisure facilities and the planting of forests should be allowed.

Many of these issues are ones on which geomorphologists could contribute from their understanding of processes and of interactions in the environment.

## PRESENT STATE OF RESEARCH

'Applied' geomorphology has been identified for a long time. However, there has been a change in attitudes and type of research particularly since about 1982—to quote one participant in the symposium on public policies 'geomorphology has become much more exciting'. Undoubtedly geomorphologists have become much more involved in practical work and have been consulted over environmental problems to a greater extent but it is still rather limited. How the chapters presented here differ from other applied work in this field is that the participants seem to have become more aware of the policy frameworks in which they are operating and have not been content simply to present technical data and hope it will be used. They have fitted their work into the framework, have provided the type of information needed and have pursued its application in terms of policy interaction.

There are a number of reasons why this change may have taken place. It may be that geomorphological studies had advanced sufficiently that the subject was ready to become involved and provide answers. Undoubtedly political pressures and economic forces in higher education and research establishments have meant that we have had to justify our existence and increasingly pay for our activities. The client wants usable information from which policy decisions can be made. There is much debate on the extent to which geomorphologists and researchers in general should do consultancy work because of its narrow objectives and constrained programmes. Most workers are agreed there is still considerable need for non-specific research and some have successfully combined the two by getting applied work to pay for some additional pure research. Rothschild's report to the Heath Government said at least 10 per cent of contracts should be reserved for strategic research but this was never followed. It does seem however that much more of geomorphological work is applicable if only other professionals and policy-makers were aware of its scope and capabilities. The methods by which this can be brought about are discussed below.

Four main types of work can be seen to interface with policies to different degrees and be the kind of research or consultancy which geomorphologists can offer or be asked to do:

(1) *Cataloguing and inventories.* This kind of work has often been derided academically as low level but information of 'what is where' is often basic to policy implementation and even design and is the kind of information which public agencies require and which can quite easily, given the resources, be provided. Provision of this type of information is exemplified by the landslides study of the UK (Chapter 7) and by Boardman and others' work on soil erosion though these also show that the work may be demanding and require full use of geomorphological skills. Similarly in the Arctic development in Canada the Canadian Government has called upon geographers and geomorphologists to provide inventories of specific types of terrain and materials.

(2) *Assessment of effects of activities.* This has been the scope of much 'pure' research in recent years, to understand the processes in general. This kind of approach can now be applied to specific locations though the work often entails monitoring for some time so quick answers cannot be obtained. However, the experience within this field is increasing and the ability to assess effects must be communicated effectively to other professionals and planners.

(3) Following closely on from the previous type of work is the ability to *predict effects of proposed activities.* This arises directly from experience of case studies and from research into predictive models. Here, some of the problems of environmental work which make it of less appeal to clients than, say, engineering are apparent. The complexity of the environment, the number of factors involved and the great variation in characteristics mean that it is very difficult to predict exact outcomes. Potential clients find this unsatisfactory and often do not want answers hedged with 'perhaps' and 'maybe' and 100 per cent margins. Perhaps geomorphologists have to be prepared to 'put their heads on the block' rather more. Also though, it is a matter of environmental education so that clients realize the nature of the environment and the kind of results which will be obtained. There are some signs that awareness of environmental complexity is increasing.

(4) The fourth type of work is that in which the geomorphologists not only provide technical data but *suggest policies and evaluate alternative policies.*

This means an awareness of the political and economic frameworks since it is only if it is aimed at and communicated to policy makers that it will have any effect. Again there has been a certain reticence to express such 'political' views but as several chapters in this volume show it is only in this way that much of the work can be effective. It is naive to think that a sound piece of scientific research will in itself produce changes in policy. Newson (Chapter 2) cites cases of scientific data being ignored or manipulated. It may be that in the end it is persuasive argument which is needed.

## SCOPE FOR GEOMORPHOLOGY

There are a number of reasons why geomorphologists, both individually and collectively, wish to become involved in policy formulation to a greater extent. These include (i) a personal concern with the state of the environment and the way it is managed, (ii) social responsibility—the desire to do work which is, and is perceived to be, socially valuable, (iii) as a means of obtaining funding and a response to political pressures to ensure survival or enable other work to be done, and (iv) to enhance the profession so that geomorphology gains greater recognition which again gives greater personal satisfaction and increases professional opportunities.

A major question is what are the strategies for involvement in public policies? It has already been pointed out that mere presentation of scientific evidence, however convincing to the scientist may be insufficient. Portions of the data which support a particular case may be selected. The evidence may be dismissed as inadequate no matter how good the research. However, there are also cases where scientific communities have had large impact even without lobbying and persuasion at the political level—this relates again to professional standing and academic clout.

If policy is to be influenced then it is important to understand who the policy makers are and who they are advised by. In many cases the concern may not be in achieving particular legislation but rather with implementation in practice. Thus many of the issues need to be tackled at local level where it is involvement with the local planners which is important.

If this type of work is to increase then geomorphologists must have a recognizable profession and should have marketable skills and techniques. There is some debate within the discipline as to whether this should entail professional 'standards', as for example in engineering. It may also mean going along the path of acquiring a 'chartered' status. Others consider that it is important to retain flexibility and that environmental work is not amenable to such an approach.

Awareness of the capabilities of geomorphology as a profession can be increased by publication in the journals of other professions and by producing material orientated to public consumption as well as by attending meetings and communicating with other professionals. Professional barriers may be difficult to overcome but there are benefits to be gained by both sides.

Another strategy is for the professional societies, representatives and individuals to become involved in the political process. This may be done by political lobbying and contacting politicians or by involvement in political parties. The essential ingredient is effective communication between geomorphologists and those needing the information or needing to be made aware of the problems and policy voids.

## CONTENTS

This volume is divided into sections according to spheres of human activity and types of geomorphological zone. In each section the present scenario is reviewed and a case study then exemplifies some of the issues and potential for geomorphological involvement. The first section deals with impact of land use policies particularly afforestation in the uplands (Newson, Chapter 2) and growth of cereals in the lowlands (Boardman, Chapter 3). The latter demonstrates the increasing problem of soil erosion in Britain and Morgan and Rickson (Chapter 4) then discuss one possible strategy for preventing or alleviating erosion.

The second section examines the impact of urban land uses. Douglas (Chapter 5) provides a review of the situation, focusing on hydrological issues and incorporating lessons to be learned from locations outside Britain. Aspinall, Macklin and Openshaw's paper (Chapter 6) is a case study of the legacy of industrial and urban activities on the soil, the consequences for environmental health and implications for policy.

Direct constructional or destructional activities can affect or be affected by the stability of slopes and the needs for information and strategies for its provision are discussed by Brunsden (Chapter 7). Gagen and Gunn's paper (Chapter 8) applies to a specific area of activity, that of limestone quarrying but produces practical recommendations on policy and practices.

Rivers and coasts are both dynamic environments which pose particular problems of management. Brookes and Gregory (Chapter 9) outline the issues relating to channelization of streams and the case for more geomorphologically-based design schemes. Crabtree's paper (Chapter 10) is concerned with a rather different aspect of river basins, that of pollution, specifically from sewer overflows, but he shows how understanding of river basin dynamics is playing a part in formulating policy.

In the coastal section all three papers demonstrate the problems caused by the dispersed nature of coastal management and number of institutions involved. Carr (Chapter 11) reviews the history of this field while Penning-Rowsell,

Thompson and Parker (Chapter 12) demonstrate the complexity of factors to be evaluated in coastal impact assessment. Sims and Ternan (Chapter 13) demonstrate the conflicts between coastal protection and planning policies and the necessity of understanding the physical situation.

Policy-makers and users of information are represented by Brook and Marker (Chapter 14) of the Department of the Environment who indicate their data and information needs and what geomorphologists need to do, particularly using the example of slope instability.

The conclusions to be drawn from these papers in terms of scope for geomorphology in environmental planning and policy are summarized by Hooke (Chapter 15).

## REFERENCES

Blunden, J. and N. Curry (eds) (1985). *The Changing Countryside*, Open University, Milton Keynes.

Briggs, D. (1986). Environmental problems and policies in the European Community. In C. C. Park (ed.) *Environmental Policies*, pp. 105–144, Croom Helm, London.

Coates, D. R. (1984). Geomorphology and public policy. In J. E. Costa and P. J. Fleisher (eds) *Developments and Applications of Geomorphology*, pp. 96–115, Springer-Verlag, Berlin.

Countryside Commission (1987). New opportunities for the countryside, *Countryside Commission Publication 224*, Cheltenham.

ESRC (1987). Environmental issues, *Newsletter 59*, pp. 9–26.

Gilg, A. (1986). Environmental policies in the United Kingdom. In C. C. Park (ed.) *Environmental Policies*, pp. 145–82, Croom Helm, London.

Open University (1985). *The Countryside Handbook*, Croom Helm, London. 98pp.

O'Riordan, T. (1981). *Environmentalism*, Pion, London. 409pp (2nd edition).

Park, C. C. (1986). Environmental policies in perspective. In C. C. Park (ed.) *Environmental Policies*, pp. 1–44, Croom Helm, London.

# Rural Land Use and Soil Erosion

Geomorphology in Environmental Planning
Edited by J. M. Hooke



# 2 Upland Land Use and Land Management—Policy and Research Aspects of the Effects on Water

**MALCOLM NEWSON**
*Professor of Physical Geography, University of Newcastle upon Tyne*

## THE BRITISH UPLANDS: PHYSICAL AND POLICY BACKGROUNDS

Few individuals or agencies will agree on a map definition of the British uplands, though most can recite a list of 'upland problems'. Perhaps a firm boundary is inappropriate since one of these problems is marginality and margins are bound to fluctuate, especially when, as in the case of the British uplands, they respond to changes of physical as well as economic climate (Parry, 1978). However, economic policy reactions to physical marginality have created a modern palimpsest of bureaucratic boundaries for the uplands and the most important of the later ones is that created by the Less Favoured Areas Directive of the European Community (EC 75/268) (see Figure 2.1). Even so, statistics are hard to come by for the British upland LFAs and their area was extended as recently as 1984.

The problem area constitutes one-third of our land surface, by no means negligible; but because it produces less than 10 per cent of our output of rural products, and is the home for more sheep and cattle than people, it has until recently been neglected in policy terms. Whilst post-war policy has concentrated on sustaining individual upland activities there is little effort, despite much intellectual debate, to secure specific regional and inter-agency effort. In this respect it is arguable that the uplands are *not* a special case in policy terms amongst the rural areas of Britain (Davidson and Wibberley, 1977). Their land use is substantially unplanned, albeit within some highly constrained 'free market forces'. Agriculture dominates upland land use and so its fiscal policies are a major influence; of 8 m hectares of upland, over 7 m ha are in agriculture. Other, minority, uses with which we deal here amount to about 0.6 m ha in the catchment area of major reservoirs (0.13 m ha owned by water authorities) and 0.6 m ha under commercial forestry (Tranter, 1978).

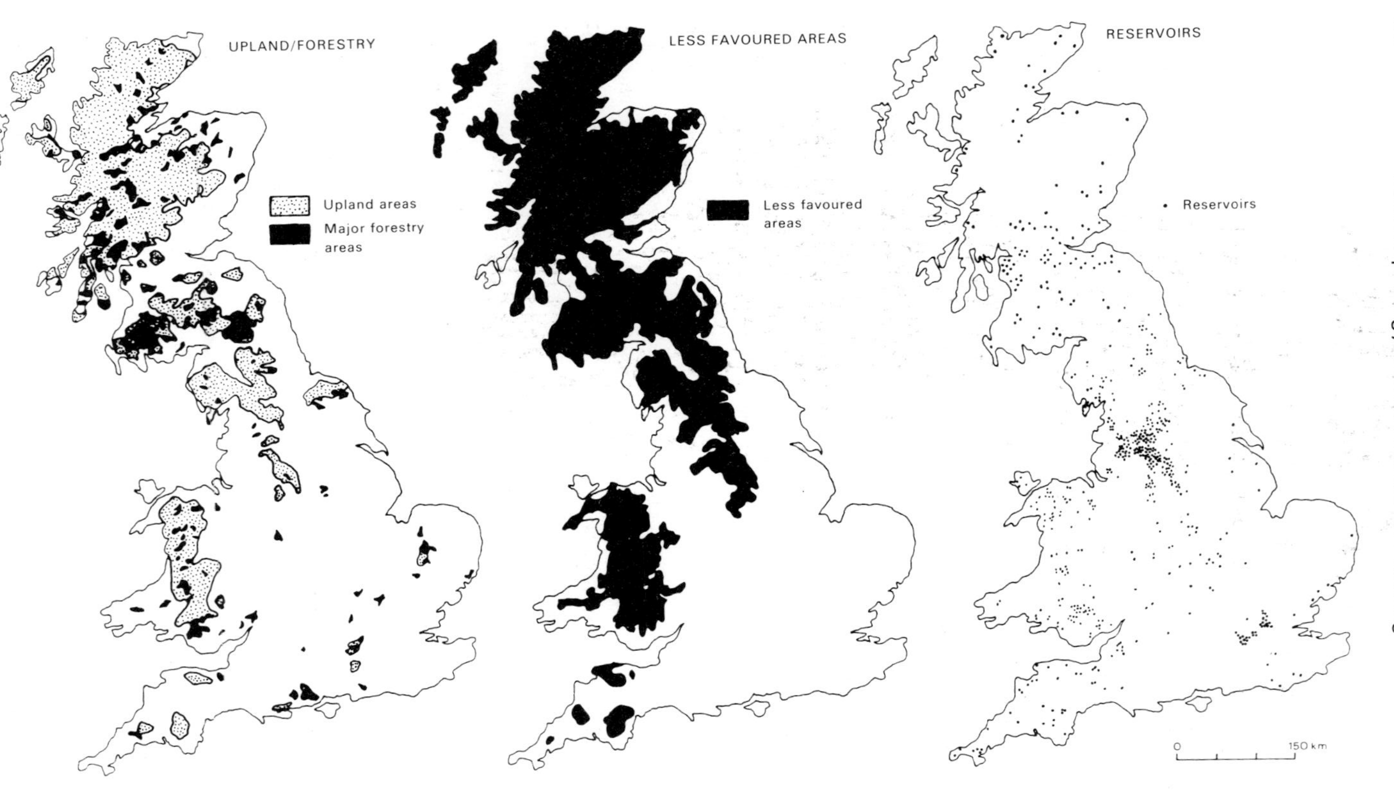


**Figure 2.1** The policy geography of the forest: water debate showing the British uplands, their forest cover, their European Community status and the location of reservoirs

The climate of the British uplands is cool, wet and windy; it acts mainly as a constraint to economic activity and richly deserves its classification as 'hyperoceanic' or 'euoceanic' (Birse, 1971; Bendelow and Hartnup, 1980). Parry (1978) has shown how the degree of physical constraint has varied in recorded history. Sporadic mineral exploitation has been possible, when the price was right, over much of the uplands. Technical innovation in machinery and crops, aided by structural and production finance has enabled pastoral agriculture to expand drastically since 1947 (see Parry, Bruce and Harkness, 1982). This era is now ending amidst a chorus of environmentalist criticism concerning the undesirable effects of (over-)productive agriculture on ecosystems (e.g. Smith, 1985). The original aim of financial support was the sustenance of rural communities; even this aspect of policy is seen to have failed (MacEwan and Sinclair, 1983).

Against this background of sporadic exploitation two current, if minor, uses of the uplands seem to have a compelling logic: the development, on behalf of lowland consumers, of water resources (because of their abundance and purity) and the plantation of exotic forests to exploit the impoverished soils (see Figure 2.1). Afforestation may also be seen as righting the wrongs of our forebears (since the Neolithic) who have delivered us into the hands of import-dependency by scant regard for timber as a renewable resource. There are thus frequent calls (e.g. Forestry Commission, 1979) for further expansion of (re-)afforestation and a tree cover is currently under consideration for the 1 m ha of agricultural land in Britain for which we are currently seeking alternative uses.

However compelling the resource arguments, forestry and water do not combine happily as joint users of the same upland areas and their conflicting interests in what amounts to a policy vacuum is the basic theme of this chapter; it has critical implications for the commissioning, direction and use of hydrological and geomorphological research. It also has a more practical aim: to sound a warning about scientific naïvety towards the processes by which society identifies, and 'solves' environmental problems by ignoring, over-reacting to, or challenging research.

## POST-WAR RESEARCH AND CHANGING ATTITUDES TO AFFORESTATION BY THE WATER INDUSTRY

If precedence were a principle and land allocation a policy in the British uplands, the arrival of water-gathering in the uplands during the Victorian era of public health improvement would give the modern water industry a powerful say in the allocation of land for commercial coniferous forestry, a need which has largely arisen since the creation of the Forestry Commission in 1919. Two major inconsistencies in the attitude to land use taken by the water industry itself, however, spoil this simple argument. Firstly, British water engineers and scientists are not accustomed to invoking catchment area processes to explain river dynamics; this is the province of the academic geographer (Newson 1987). Secondly, and more specifically, for at least forty years the water industry gave a cautious *welcome* to conifer plantations on catchment areas. Newson (1986a) reviews the history of attitudes within the water industry and how research results, principally those of Law (1956), eventually impinged on decision-making.

Law's results, which 'proved' that trees 'use' more water than rough moorland, did not lead to a change of policy but to an intensification of research. Indeed there was no public policy to change, catchment area land use being mainly decided by individual water suppliers, there being a desire to own whole catchment areas in order to prevent public access and agricultural improvement. More than twenty years later the results of Calder and Newson (1979) were met with less criticism but eventually had the same effect, that of promoting yet more research but no changes in policy.

Clearly, the reaction time between research results and policy changes is a long one, especially if there is no specific policy on which to build. By the 1970s the ability of the water industry to control catchment land use by purchase was much diminished and development of a public policy driven by research results was by no means irrelevant. It is interesting to speculate about a much faster policy reaction had the emerging resource conflict occurred in an urban context.

A detailed chronology of events surrounding the conflict in the late 1970s points up further salient features of the relationship between publication of research results and the development of public policy. In September 1977 the Centre for Agricultural Strategy (CAS) at the University of Reading held a symposium on 'The Future of Upland Britain' (Tranter, 1978). Water resources topics provided 10 per cent of the input to this potentially influential forum. Only the contribution from Devenay (1978) mentioned land use conflict with forestry, '. . . because afforestation reduces runoff which in turn affects the water supplied by a given reservoir or intake'. Baldwin (1978) contributed the opposite view. 'The Forestry Commission strives continuously to increase afforestation in upland Britain. In the author's opinion this activity is wholly beneficial to water supplies interests'. Shearer (1978) was even deferential to forestry, writing of the proposal to create the Kielder Reservoir that, 'the land is heavily wooded with conifer . . . the Forestry Commission . . . do not object to the proposal'. The symposium took the view, therefore, that forestry and water were compatible harvests from our bleak uplands.

Eighteen months after this first CAS discussion of forestry and water, the Centre hosted a meeting to discuss a draft 'Strategy for the UK Forest Industry' (CAS, 1980). A threshold change had occurred in the water industry's view, now officially represented by the Central Water Planning Unit, which had used the results from the Institute of Hydrology on water yield (see Calder and Newson, 1979) to predict that extra costs of up to £2m could be incurred if afforestation took place on the catchment areas of three sample upland reservoirs (Figure 2.2).

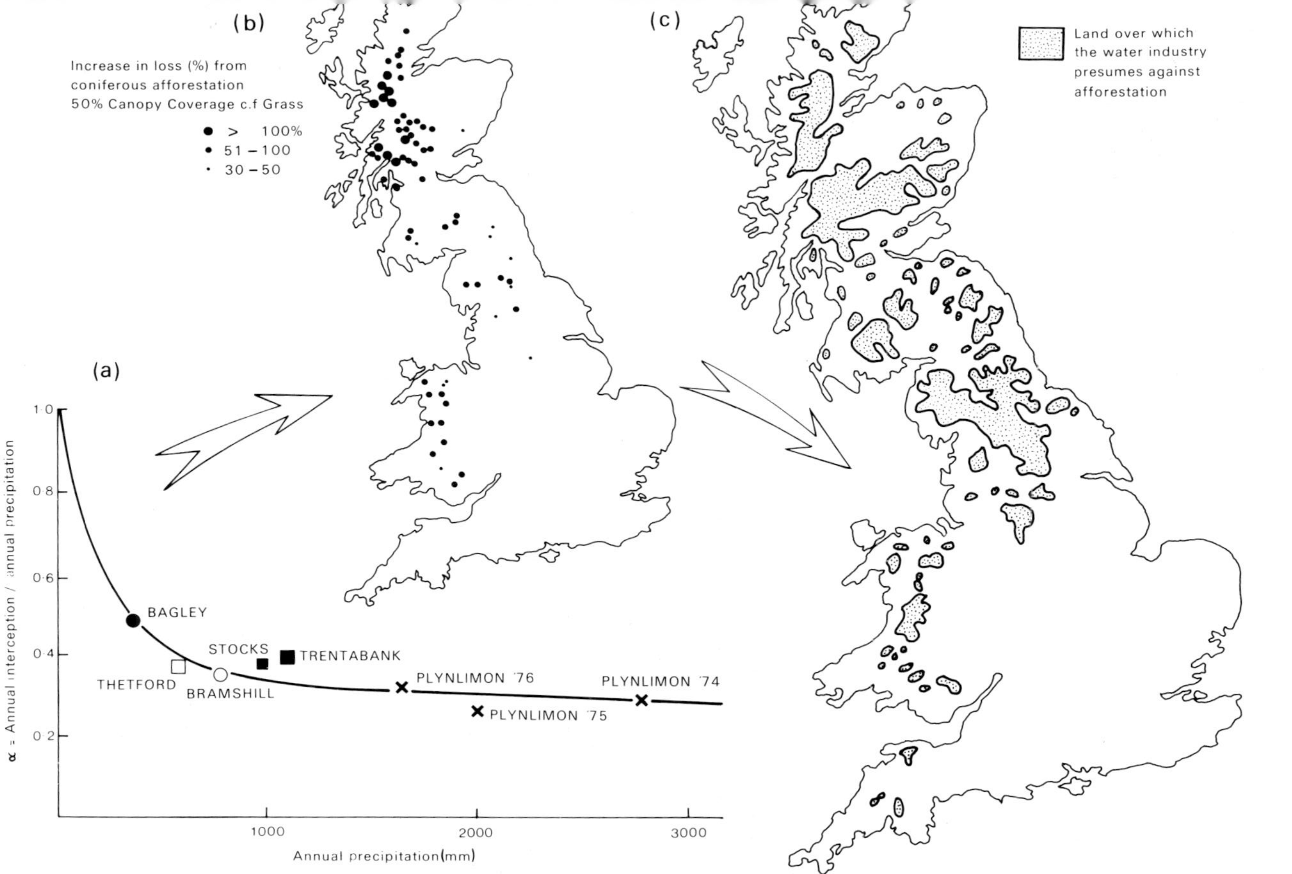


**Figure 2.2** Aspects of the science: policy debate. Empirical results (on interception ratio of forests) produce model predictions which lead to policy viewpoints

This financial line of argument was thoroughly in the tradition of Law (1956), used to create impact but not sufficiently amenable to consensus to be a basis for rational resource allocation. The report of the CAS exercise included a map of land suitable for afforestation in Britain against a background of constraints including, as the most important single category, water catchment areas (Figure 2.2). It is interesting to note that water quality considerations were at this stage mere adjuncts to the main area of contention: water yields. The era was one in which the demand for water looked set to increase by 45 per cent by the end of the century, the industry had no desire to get into further political battles to build more dams in upland valleys, and the upturn in public perception of water quality, supported by European Community Directives, was not foreseen. Between the researchers and technicians of the forest and water industries debate had reached a high intensity. However, the distance between the technical situation (a virtual 'keep off' message from catchment managers to foresters) and the public policy arena is summed up by a statement from Mr Tom King to Parliament (*Hansard*, 21 March 1980) to the effect that, 'As regards afforestation, its percentage and its effect on catchment areas . . . I am advised that there is lack of clear scientific evidence'. Deliberate inertia in public policy formulation often sustains itself with assertions that there is no evidence of a clear policy issue. What is often meant, however, is that research results are still required to pass through the political filter between science and policy before emerging as an issue. Crises and catastrophes speed this process but these were not a feature of the upland land use debate in relation to water. In fact there were other issues, of nature conservation, water quality and fisheries which were to bring much greater pressure on the policy makers with, arguably, a weaker initial factual base.

## SIGNALS FROM THE MOUNTAIN STREAM OF THE NEED FOR POLICIES

Technical debate over the relationship between forestry and water supply once again proved creative for research. The outlook for timber production (Forestry Commission, 1979) suggested that Scotland would bear the brunt of new planting. Already the Scottish local councils and hydroelectric boards were identifying local problems in connection with afforestation. Since research results were not available in Britain on the effects of a 'natural' vegetation dominated by heather or of frequent snowfall (both conditions likely to make Scottish afforestation unique in its hydrological effect) the Institute of Hydrology set up a paired catchment study on the Plynlimon model at Balquhidder, Perthshire. Work began in 1980.

By now the water quality dimension was becoming important to water industry perceptions of land use and land management (see Youngman and Lack, 1981).

Issues not strictly related to water supply, such as fisheries, also began to surface (Harriman, 1978; Harriman and Morrison, 1982) and none of these research results was favourable to forestry. The Institute of Hydrology at Plynlimon was also turning to research on water quality, stimulated by the results of Stoner, Gee and Wade (1984) who reported a fisheries decline associated with forests in Wales. Much of the expanded research effort took a catchment scale and few resources were available for process studies, a dangerous situation, in that, without calibration of causative processes, there is no basis for extrapolation of significant results (Ward, 1971). This research effort was also criticized by Kay, McDonald and Jenkins (1982) for being subjectively anti-forestry in approach, though upland agriculture's boom was also coming under water quality scrutiny by the mid-1980s (Roberts, Hudson and Blackie, 1984, 1986).

Results are now emerging from the expansion of research in the 1980s. What signals do they give for movement in the field of public policy?

## Water Yield

Evaporative losses from upland catchments covered by conifer plantations are greater than those from short-grass moorland. The effect is predictable using a simple model (Calder and Newson, 1979) because the interception process, which explains the finding, is controlled by crop height and climatic conditions. Increases of loss up to 136 per cent are predicted in high rainfall areas (Figure 2.2) with extensive forest cover; reductions of runoff are less spectacular, up to 20 per cent. At a time of little growth in the demand for water the impact of this result is hardly severe, except where energy costs are incurred by the need for extra pumping in conjunctive-use schemes. The cost of energy also makes reduced annual runoff important for hydro-power generation and an economic analysis based on energy units (Barrow, Hinsley and Price, 1986) is not favourable to afforestation at high altitudes on HEP catchments.

## Runoff Extremes

Interception losses create soil storage for moisture beneath mature conifers which may reduce runoff from small to moderate rain storms (Robinson and Newson, 1986). However, during the early stages of afforestation, where ground preparation involves ploughing and drainage, storm runoff is more rapid and higher flood peaks result, at least locally (Robinson, 1980). These effects persist in the more densely drained plantations into the mature crop (Robinson and Newson 1986).

In droughts, control of runoff passes from the vegetation and soil to aquifers beneath: blanket peat, drift and fractured bedrock in the uplands. These have relatively little storage capacity and the reduction of recharge by afforestation may not affect minimum flows. However, small sources used for rural water supply may experience earlier and more sustained drought if recharge areas are completely afforested, especially where drainage directs recharge water away from the intake works. Similarly, interception and increased soil moisture deficits will reduce the effectiveness of 'recovery rains' for all catchments during and after droughts. Reservoirs with afforested catchments recover more slowly after droughts.

### Acidity and Metals

Most upland streams give an acid reaction (Newson, 1986b). Streams draining forested catchments on base-poor rocks and drifts produce notably more acid runoff than their moorland counterparts. Stoner and Gee (1985) illustrate the effects of reduced pH on the solubility of soil metals; this raises, for example, the aluminium content of streams and it may well be via metal toxicity that fisheries are diminished beneath a conifer cover.

### Sediment Yields

In areas where non-cohesive soils and drifts are ploughed and drained to establish forests there is a considerable increase in the sediment load delivered to streams (Moore and Newson, 1986; Stott *et al.* 1986). Deterioration of reservoir water quality can occur (Austin and Brown, 1982; Stretton, 1984). The increase in sediment yields persists through the life of some forest crops and there is considerable risk of further disturbance during timber harvesting. Despite care in planning ground-trafficking during felling, the design and construction of forest roads to permit an efficient scale of timber removal seems destined to result in high, if transient, sediment loads.

### Water Temperature

In the first rotation of British upland conifer crops it was common to plant right up to channel banks. Lack of light penetration to the stream once the trees mature clearly cools the water and inhibits feeding and breeding by fish (Smith, 1980).

### Nutrients, Bacteria, Herbicides and Pesticides

These parameters are not the subject of detailed research although each has contributed to the technical debate about land use and land management. For example, Harriman (1978) pointed to the danger of eutrophication in Scottish lochs following routine fertilizer applications to afforested areas in their catchment area. At Plynlimon, however, the attention of the Institute of Hydrology was drawn more to similar dangers from agricultural fertilizer applications (Roberts, Hudson and Blackie, 1984). Kay, McDonald and Jenkins (1982) point to the bacteria purity of forest drainage waters, indicating a significant advantage to the water industry over improved pastures with densely grazed livestock, from which there is a distinct threat of bacterial pollution to recreational waters (McDonald, Kay and Jenkins, 1982).

From time to time herbicides and pesticides are misused in forestry plantations, resulting in local pollution incidents, but these and the effects of the other parameters in this group are amenable to control by water pollution legislation, by recommendations on the use of water, or chemicals, and by changes in management such as ground applications of aerial sprays. In other words, results in this category do not bespeak changes in policy affecting land allocations because there are alternative ways of managing the issues they raise. This example therefore illustrates the value of awareness of relevant policy issues in research design: research on fertilizer effects should crucially examine application rates, timing and downstream dilution.

## THE FUTURE: LAND USE ALLOCATION OR LAND MANAGEMENT ACCOMMODATION?

It is useful to maintain and develop the dichotomy between issues of land allocation and issues capable of resolution through existing legal or technical fixes, allowing a harvest of both water and timber from the same land in the uplands. Given present suggestions that an increase in plantation forestry may be an answer to the problems of land not needed for farming (because of agricultural surpluses), we may set up two scenarios for decision-making:

### The *Allocation* Option

A 'keep-off' attitude to catchment areas may be followed by the water industry, armed with maps of sensitive areas such as that published by CAS (1980) and that used by Welsh Water. A refinement of this approach might be to plan catchment land use rationally on the basis of land capability assessments and hydrological predictions, but, as Newson (1984) shows, this approach is at times strapped for data on both land capability and water quality. A more serious problem is that this approach cannot make logical inputs to a policy framework because none exists!

### The *Accommodation* Option

Here land is allocated by a combination of 'free' market forces and a technical dialogue between the forest and water industries. Both water and timber are harvested from the same land but both industries accept that higher costs may be involved, e.g. for greater care in preparing ground for afforestation, for leaving large strips of land unplanted or for higher levels of water treatment from upland sources.

In practice a third 'middle way' is more likely, the precise balance between options being determined by local factors, not the least of which is pressure from outside interests using their own formal or informal means of applying pressure, e.g. conservation and recreation. The 'middle way' is also more likely because water quality issues now dominate the debate and because the debate is going on in the UK. As Hawkins (1984) emphatically concludes, enforcement of environmental pollution laws and directives in the UK is an exercise in the art of the possible. Thus 'keep off' attitudes will only prevail for very sensitive sites or for persistent and, if it can be proved, deliberate contributions to the deterioration of upland water quality. Standards and codes of practice are needed at this stage; at present only very provisional ones exist (e.g. Mills, 1980; Timber Growers UK, 1985).

Even the widespread adoption of the 'middle way' bespeaks two developments of which there are few signs at present:

(1) The creation of a formal structure for dialogue between the forest and water industries; *ad hoc* structures involving the relevant statutory bodies exist in some places, largely as the result of pressure from the other interest groups, e.g. in Wales, but elsewhere attitudes can be as entrenched as in the days of Law's experiments.
(2) The creation of a body of practical research results which can form the basis of allocation or accommodation strategies for individual sites. At present only 'broad-brush' statements exist and there are great dangers (see below) when scientific conclusions are precipitously poised on the brink of application. Not the least of these dangers is the across-the-board adoption of, say, contour ploughing in forestry land preparation, because it appears to 'work' in one or two sites. There are unlikely to be such miracle conclusions and 'Do your best' solutions will abound.

## RESEARCH AT THE INTERFACE WITH POLICY: A CAUTIONARY NOTE

Investigating the relationships between land use, land management and water resources has a compelling attraction as an example of applied geography. Experimentation of the type referred to earlier gains distinction for the physical geographer as a bona fide scientist, whilst the context of the work includes historical legacies, resource evaluation, economics, planning and politics, and thus is potentially open to the full width of geographical expertise. However, the analysis presented above points out some pitfalls which the geographer reader might take care to avoid. Clark (1977) has prepared the geomorphologist working on coasts for frustration, as solutions to problems yielded by research are rejected or damned-by-faint-praise at the decision-making stage. Such frustrations can seldom be avoided but the learning curve of political reality can be made shallower if it commences earlier. Researchers should check the policy context of their work before setting out, particularly in the present economic climate in which we are all encouraged/forced to accept financial inputs from the participants in policy formulation (Johnston, 1981).

The salient features of the *present* state of the upland management debate may serve to help us gain a defined awareness of, and therefore a professional approach to, public policy studies. The first lesson is clearly that contexts change rapidly; by the time this chapter is published there may be an entirely new policy framework for the uplands. The other lessons are appropriate to stages along the line of policy formulation. At the stage where there is *no policy*, for example on rural land-use planning (Davidson and Wibberley, 1977) there will be considerable rejection of research results. O'Riordan (1976) suggests that 'Where problems pose solutions which challenge the dominant values and rules of political consensus, substantial power may be directed simply at keeping this challenge out of the political arena'. Under such circumstances the researcher may well feel jealous of the effect evoked by qualitative rather than quantitative evidence, such as in the case of the high impact of the conservation lobby in the uplands (see recent publications by Nature Conservancy Council (1986) and Tomkins (1986)). O'Riordan (1976) finds society's fears of rationality predictable; as a direct result, 'policy making is basically a political process'. Thus Smith (1984) concludes in gloomy terms about the potential for applying a computer model for planning upland land use. Mather (1986) warns that land use planning, 'cannot be simply a technical, value-free exercise, although it has technical aspects'.

At a second stage one can detect that society 'feels *a policy coming on*'. At present in the uplands the issue at the centre of this chapter is likely to become a special case within a much more formal interventionist structure. Researchers may well feel resentment that moves towards policy will be led by other issues, principally nature conservation and agricultural production. Thus, once again, numerical inputs to a rational model for land use are swept aside by fiscal, social and even ideological considerations. In the uplands the activities of single-interest agencies clearly need to become broadened and co-ordinated whilst the value of land is manipulated to achieve some form of planning. Roome (1984) concludes, there must be, 'fundamental change to the distribution of rights, whether voluntarily accepted or legally enforced'. Clearly we are some distance now, both conceptually and methodologically, from the interception of rainfall!

A third-stage can now be glimpsed. With the exception of the Conservatives (despite exhortations from the Bow Group), all other parties produced manifestos for the 1987 General Election calling for an *increase in planning controls* for rural land. The SDP's view was worthy of quotation here, 'In determining environmental policy due weight must be put on the evidence produced by scientific research'. Following the election outcome we probably cannot anticipate an interface between research and policy which is free of the frustrations described above, and which results from a thorough policy for the environment (including land use). Evidence in the environmental sciences is often likely to be inconclusive, by the scientist's own high standards, and it is likely that evidence will be judged by the way in which it serves existing policy. Collingridge and Reeve (1986) describe 'an unhappy marriage' between science and policy. They describe the 'myth of rationality' behind which incremental accommodations between protagonists inveigles researchers to produce results which only then feed controversy standing in the way of fundamental progress.

Clearly the message is that we should give as much attention to the public policies (or lack of them) as we give to the geomorphology. Even so having mastered the whole picture, we should not be arrogant enough to expect our careful, studious rationality to be matched by those who use our work and, increasingly, sponsor it. Science will remain a servant to society, not its master.

### Acknowledgements

The author acknowledges the experience of upland issues created by thirteen years at the Institute of Hydrology's Plynlimon office. He is grateful to a new colleague, Martin Whitby, for helpful comments and to an unknown reviewer, commissioned by the editor, for guidance.

## REFERENCES

Austin, R. and D. Brown (1982). Solids contamination resulting from drainage works in an upland catchment and its removal by flotation, *Jnl. Instn. wat. Engrs. & Sci.*, **36**(4), 281–8.

Baldwin, A. B. (1978). Quality aspects of water in upland Britain. In R. B. Tranter (ed.) *The Future of Upland Britain*, Centre for Agricultural Strategy, Reading. pp. 322–7.

Barrow, P., A. P. Hinsley and C. Price (1986). The effect of afforestation on hydroelectricity generation, *Land Use Policy*, 141–52.

Bendelow, V. C. and R. Hartnup (1980). Climatic classification of England and Wales, *Soil Survey Tech. Mono.* 15. Harpenden.

Birse, E. L. (1971). *Assessment of Climatic Conditions in Scotland.* Soil Survey of Scotland, Aberdeen.

Calder, I. R. and M. D. Newson (1979). Land use and upland water resources in Britain, a strategic look, *Water Res. Bull.*, **15**(6), 1628–39.

Centre for Agricultural Strategy (1980). Strategy for the UK forest industry, *CAS Report* 6, 347pp.

Clark, M. J. (1977). The relationship between coastal zone management and offshore economic development, *Marit. Pol. Mgmt.*, **4**, 431–49.

Collingridge, D. and C., Reeve (1986). *Science Speaks to Power*, Frances Pinter, London.

Davidson, J. and G. Wibberley (1977). *Planning and the Rural Environment*, Pergamon Press, Oxford.
Devenay, W. T. (1978) Water supply in upland Scotland. In R. B. Tranter (ed.), *The Future of Upland Britain*, Centre for Agricultural Strategy, Reading. pp. 328–35.
Forestry Commission (1979). *The Wood Production Outlook in Britain*, Forestry Commission, Edinburgh.
Harriman, R. (1978). Nutrient leaching from fertilised forest watersheds in Scotland, *Jnl. appl. Ecol.*, **15**, 933–42.
Harriman, R. and B. R. S. Morrison (1982). Ecology of streams draining forested and non-forested catchments in an area of central Scotland subject to acid precipitation, *Hydrobiologia*, **88**, 251–63.
Hawkins, K. (1984). *Environment and Enforcement*, Oxford University Press.
Johnston, R. J. (1981). Applied Geography, quantitative analysis and ideology, *Appl. Geog.*, **1**, 213–19.
Kay, D., A. McDonald and A. Jenkins (1982). *Upland Catchment Research: an Integrated Perspective*. Paper presented at the British Geomorphological Research Group Annual Conference, University of Birmingham.
Law, F. (1956). The effect of afforestation upon the yield of water catchment areas. *Journal of British Waterworks Association*, 38, 484–94.
MacEwan, M. and G. Sinclair (1983). *New Life for the Hills*, Council for National Parks, London.
Mather, A. S. (1986). *Land Use*, Longman, Harlow.
McDonald, A., D. Kay and A. Jenkins (1982). Generation of fecal and total coliform surges by stream flow manipulation in the absence of normal hydrometeorological stimuli, *Appl. and Eng. Microbiol.* **44**(2), 292–300.
Mills, D. H. (1980). The management of forest streams, *Forestry Commission Leaflet*, 78, HMSO, London.
Moore, R. J. and M. D. Newson (1986). Production, storage and output of coarse upland sediments: natural and artificial influences as revealed by research catchment studies, *J. Geol. Soc.*, **143**, 921–6.
Nature Conservancy Council (1986). *Nature Conservation and Afforestation in Britain*, NCC, Peterborough.
Newson, M. D. (1984). Upland reservoir catchments: towards multiple resource planning? *Cambria*, **11**(2), 73–86.
Newson, M. D. (1986a). Concepts of management—surface water: the example of the British uplands. In F. T. Last, M. C. B. Hotz and B. G. Bell (eds), *Land and Its Uses, Actual and Potential*, Plenum Press, New York. pp. 179–92.
Newson, M. D. (1986b). Slope and channel processes in upland catchments: interfaces between precipitation and streamflow acidity. *Cambria*, **13**(2), 197–212.
Newson, M. D. (1987). Land and water: The 'river look' on the face of geography, *Univ. of Newcastle Geography Dept. Seminar Paper* 51, 32pp.
O'Riordan, T. (1976). Policy making and environmental management: some thoughts on purposes and research issues, *Nat. Res. Jnl.*, **16**, 55–72.
Parry, M. L. (1978). *Climatic Change, Agriculture and Settlement*, Dawson, Folkestone.
Parry, M. L., A. Bruce and C. E. Harkness (1982). *Moorland Change Project*, Department of Geography, University of Birmingham.
Roberts, G., J. A. Hudson and J. R. Blackie (1984). Nutrient inputs and outputs in a forested and grassland catchment at Plynlimon, mid-Wales, *Agric. wat. Mgmt*, **9**, 177–91.
Roberts, G., J. A. Hudson and J. R. Blackie (1986). The effects of grassland improvement on the quality of upland streamflow—a natural lysimeter and a small plot experiment on Plynlimon, mid-Wales, *Institute of Hydrology Report*, 96, Wallingford.

Robinson, M. (1980). The effect of pre-afforestation drainage on the streamflow and water quality of a small upland catchment. *Institute of Hydrology Report*, 73, Wallingford.

Robinson, M. and M. D. Newson (1986). Comparison of forest and moorland hydrology in an upland area with peat soils. *Intl. Peat. Jnl.*, **1**, 49–68.

Roome, N. (1984). A better future for the uplands—a planning critique, *Planning Outlook*, **27**(1), 12–17.

Shearer, D. M. (1978). Water resource development in upland Britain. In R. B. Tranter (ed.) *The Future of Upland Britain*, Centre for Agricultural Strategy, Reading, pp. 294–306.

Smith, B. D. (1980). The effects of afforestation on the trout of a small stream in southern Scotland, *Fish Mgmt.*, **11**(2), 39–58.

Smith, M. (1985). *Agriculture and Nature Conservation in Conflict—the Less Favoured Areas of France and the UK.* Arkleton Trust, Langholm, Dumfriesshire.

Smith, R. (1984). The use of computer models in planning upland land use, *Planning Outlook*, **27**(1), 18–22.

Stoner, J. H., A. S. Gee and K. R. Wade (1984). The effects of acidification on the ecology of streams in the upper Tywi catchment in west Wales, *Env. Poll.*, A, **35**, 125–57.

Stoner, J. H. and A. S. Gee (1985). Effects of forestry on water quality and fish in Welsh rivers and lakes, *Jnl. Instn. wat. Engrs. & Sci.*, **39**(1), 27–45.

Stott, T. A., R. I. Ferguson, R. C. Johnson and M. D. Newson (1986). Sediment budgets in forested and unforested basins in upland Scotland. In R. F. Hadley (ed.), *Drainage Basin Sediment Delivery*, IAHS Publ. 159, pp. 57–68.

Stretton, C. (1984). Water supply and forestry—conflict of interest: Cray Reservoir, a case study . *Jnl. Instn. wat. Engr. & Sci.*, **38**(4), 323–30.

Tomkins, S. C. (1986). *The Theft of the Hills: Afforestation in Scotland*, Ramblers Assn, London.

Timber Growers UK (1985). *The Forestry and Woodland Guide*, TGUK, Knightsbridge, London.

Tranter, R. B. (ed.) (1978). *The Future of Upland Britain.* Centre for Agricultural Strategy, Reading.

Ward, R. C. (1971). Small watershed experiments. An appraisal of concepts and research developments, *University of Hull, Occ. Papers in Geography*, 18.

Youngman, R. E. and Lack, J. (1981). New problems with upland waters, *Water Services*, **85**, 13–14.

Geomorphology in Environmental Planning
Edited by J. M. Hooke



# 3 Public Policy and Soil Erosion in Britain

**JOHN BOARDMAN**
*Department of Humanities and Countryside Research Unit, Brighton Polytechnic*

## INTRODUCTION

The interaction between erosion and public policy has long been recognized particularly in the domain of the coastal zone. Increasing population pressure on a small island and technological developments that enable large-scale changes to be made to the landscape, have operated to bring more people into contact with erosion. In some situations of conflict erosion is perceived as desirable by one interest group and is deplored by others (e.g. Clark, 1976). In agriculture, erosion is a legitimate concern of public policy because it is a threat to productivity and an expense to be borne by communities adjacent to farming land.

In Britain at present there is growing concern about erosion of upland grazing land (Mather, 1983) and peat moorland, debate about the latter topic having persisted over the last thirty years (Evans and Cook, 1986) with concern now being expressed by water and planning authorities (Phillips, Yalden and Tallis, 1981; Tallis and Yalden, 1983). In the lowlands, there is continuing awareness of the threat of wind erosion on sandy and peaty soils in the Vale of York, the Fens, the Suffolk sandlands, the Lancashire mosses and parts of Nottinghamshire but little recent data on the scale of the problem or the costs involved. In the last fifteen years water erosion has come to be seen as the major erosion problem on agricultural land in the lowlands. This chapter aims to assess the scale of the water erosion problem on the lowlands and the implications for public policy particularly with regard to the development of a soil conservation strategy for Britain.

## IS THERE AN EROSION PROBLEM IN BRITAIN?

There is little doubt that there has been a large increase in recorded erosion events in Britain in the last fifteen years (Figure 3.1). Awareness of the problem accounts for some of the increase but to offset that many instances are still

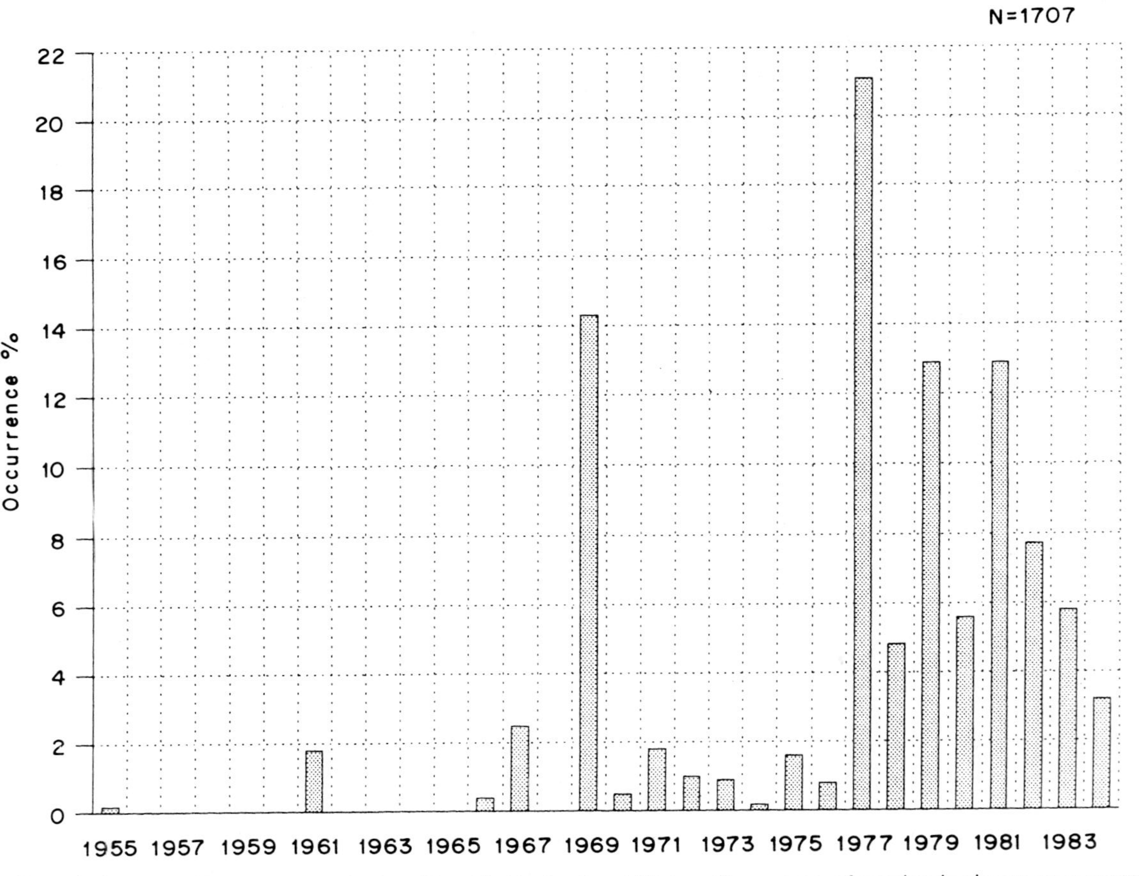


**Figure 3.1** Recorded cases of erosion on agricultural land in England and Wales. Occurrence of erosion is shown as a percentage of the total number recorded from 1955 to 1984 (from Evans and Cook, 1986)

going unrecorded because large parts of the country are not regularly surveyed. Systematic recording has only been carried out in limited areas: Reed in the West Midlands from 1967 to 1976; Colborne and Staines in Somerset and Dorset from 1982 to 1984; Boardman on the South Downs from 1982 to 1987 and Evans, for the Soil Survey of England and Wales/Ministry of Agriculture, Fisheries and Food (SSEW/MAFF) air photographic survey of seventeen sample areas from 1982 to 1986, a survey that is continuing (R. Evans, personal communication). Speirs and Frost (1987) also suggest, based on long experience of eastern Scotland, that a real increase in erosion has occurred.

There is general agreement that a large proportion of the increase is due to the adoption of winter cereals and the consequent expansion of the area of bare ground in the autumn and winter months (Boardman and Robinson, 1985; Evans and Cook, 1986; Speirs and Frost, 1987). Evans (1985) points out that the area sown to winter cereals has increased by more than three times since 1969. Crops such as sugar beet, potatoes and market garden produce are also prone to erosion but acreages under these crops are much less than those sown to cereals. Other factors that have contributed to the increase in erosion are the adoption of smoother seedbeds (Speirs and Frost, 1987), the enlargement of fields (Evans and Cook, 1986) and the occurrence of downslope wheelings in many fields (Reed, 1983, 1986). The role of a decrease in organic matter levels (Hodges and Arden-Clarke, 1986) is controversial and the necessity for high-intensity rainfall is unproven (Boardman and Spivey, in press), although it is clear that on some soils low-intensity falls can be highly erosive (Reed, 1979).

The existing database of eroded sites is incomplete and probably underestimates the scale and extent of the problem. However, it can be used to define soils at risk and these are identified in the legend to the National Soil Map (Mackney *et al.*, 1983) although notable omissions occur for which evidence has emerged since 1983. In the legend some soil associations are identified as having a risk, or a slight risk, of erosion by water or wind. However, the risk assessment applies to the land use typically found on the soil association when mapped. This may, of course, change in the future, for example, the Malham association soils of the Pennines are not regarded as 'at risk' because they are generally under grass. If a change to arable usage were to occur the soils would become highly susceptible to erosion (Boardman and Spivey, in press). The legend indicates that 44 per cent of the arable soils of England and Wales are at risk of water or wind erosion (ENDS, 1984). The 'at risk' assessment applies to soils under current land use and erosion therefore results from mismanagement or inappropriate use. The National Soil Map has been used as the basis for producing a risk assessment map by combining soil information with rainfall data (Morgan, 1985), but the selection of a valid rainfall erosivity index is open to criticism (Speirs and Frost, 1987). At the present time the National Soil Map designation of soils at risk is the best available countrywide survey of erosion hazard.

The National Soil Map and a number of papers published in the last fifteen years make clear that the erosion problem is regional and that problem areas include:

(1) The Lower Greensand soils of southern England and the Isle of Wight;
(2) Sandy and loamy soils in Nottinghamshire and the West Midlands;
(3) Sandy and loamy soils in Somerset and Dorset;
(4) Parts of East Anglia;
(5) Chalky soils on the South Downs and elsewhere;
(6) Parts of eastern Scotland.

Adequate data for an assessment of the scale of the problem have yet to be published. Most data refer to catastrophic events (Reed, 1979; Evans and Nortcliff, 1978; Boardman, 1983a; Boardman and Robinson, 1985) and systematic surveys, for example, by regularly repeated traverses across the landscape are needed (Colborne and Staines, 1985).

The SSEW/MAFF annual aerial photographic survey of seventeen sample strips has been carried out since 1982 and some data have been published which allows comparison to be made between areas (Evans and Cook, 1986; Harris, Evans and Boardman in preparation). This ongoing survey is the major data source for decision-making at a policy level. Sample areas were selected on the basis that an erosion problem was known or suspected. It is not a random sample because of the high cost of such an approach. Each strip is photographed in early summer and sites of suspected erosion are identified on the photographs. These sites are then checked in the field and measurements made to determine amounts of erosion as well as recording crop type and farming practice (Dr R. Evans, personal communication). Comparisons of results with the more systematic mapping by Reed (1979) show that the survey may underestimate by an order of magnitude the amount of erosion which occurred. However, in terms of national assessment of the erosion problem, and bearing in mind the lack of data prior to 1981, the air photographic survey is the most cost effective method of monitoring water erosion (Evans and Cook, 1986). *Ad hoc* surveys, e.g. Soilwatch (Soil Association, 1986) may supply useful information on areas for which there are no data and areas where changes in the incidence of erosion are occurring.

## SOIL CONSERVATION IN BRITAIN

During the last twenty years there has been little public recognition that an erosion problem, either regional or national, exists in Britain. Reed (1979) notes the 'rather complacent approach to the question of soil erosion in the United Kingdom'. In particular, those associated with the agriculture industry either denied that a problem exists (Hudson, 1967) or suggested that it is minimal (Davies, Eagle and Finney, 1972). A later edition of the same book continues to maintain that 'there is no serious risk of erosion except on steep slopes' and, despite evidence to the contrary, that in large sloping fields there is little risk in 'growing winter crops which protect the soil' (Davies, Eagle and Finney, 1982, p. 260). The same authors conclude that since the British climate is unlikely to change radically 'the risk of water erosion is not likely to get worse in the future'. However, this statement is not supported by the available data (cf. Figure 3.1). Prior to the late 1970s when there was a dearth of data these attitudes were defensible. Another recent comment also suggests that although there may be an erosion problem, conservation aspects have been ignored by academic geographers who have largely investigated it:

> The geographers appear to be interested only in the soil loss aspect and not in soil conservation and they can provide the ammunition for yet another assault on the farming industry by the so-called environmentalists (Spoor, 1986).

In fact, it is remarkable that in a series of studies of erosion, conservation measures have almost always been proposed.

A recognition of the need for soil conservation began with Evans (1971) who discussed some problem areas, outlined the need for further investigation and pointed out that conservation of the environment should include soil. Subsequently, conservation measures have been proposed by other authors (Table 3.1). The variety of listed measures reflects changing conditions in different parts of the UK, for instance, Somerset and Dorset (Colborne and Staines, 1985), the West Midlands (Reed, 1979) and the South Downs (Boardman, 1984). The view has also been expressed that appropriate conservation measures for the British situation have not been developed, the proof being that farmers have been reluctant to adopt them: the development of suitable techniques is therefore suggested as a priority area of research (Morgan, 1980, 1986).

None of the conservation advice listed in Table 3.1 constitutes 'public policy' since it is offered by researchers and, with the exception of Davies, Eagle and Finney (1982), published in academic journals. The one document that could be considered a policy document is a MAFF publication (Evans, 1984). This short, well illustrated pamphlet attempts to make farmers aware of the risk of erosion and advises how to avoid or control it. Unfortunately many farmers are unaware of its existence and need not, of course, follow the advice. Despite stated ADAS policy that advice on conservation issues would be available free to the farmer, the decision was taken in 1988 to charge for the pamphlet. However, it is a useful first step without which MAFF policy on soil conservation would be difficult to discern.

**Table 3.1** Conservation measures proposed by various authors

| | Measures | | | | | | | | |
|---|---|---|---|---|---|---|---|---|---|
| Reference | 1 | 2 | 3 | 4 | 5 | 6 | 7 | 8 | 9 |
| Boardman (1983a) | + | | | | + | | + | | |
| Boardman (1983b) | | | | | + | | | + | |
| Boardman (1984) | + | + | | | | | + | | |
| Boardman and Hazelden (1986) | + | | | | | | | | |
| Colborne and Staines (1985) | | | | + | + | + | + | | |
| Colborne and Staines (1986) | | | | + | | + | + | | |
| Davies, Eagle and Finney (1982) | + | | | | | + | + | + | + |
| Evans (1985) | + | + | + | + | + | | | | |
| Evans and Nortcliff (1978) | + | | | | + | | | | |
| Evans and Skinner (1987) | | + | | + | | | | | |
| Frost and Speirs (1984) | | + | | | | | + | | |
| Fullen (1985) | | + | + | + | + | + | | | |
| Pidgeon and Ragg (1979) | | | + | | | | | | |
| Reed (1979) | | | + | + | | | + | | |
| Reed (1986) | | | + | + | | | + | | |
| Speirs and Frost (1985) | | | + | | + | | | | + |
| Speirs and Frost (1987) | | | | | | | | | + |
| Stammers and Boardman (1984) | + | + | | | + | | + | | |

Key to conservation measures

1. Reduction in size of fields and length of slopes;
2. Alteration in timing of cultivations;
3. Adoption of no-till or minimum tillage techniques;
4. Avoidance of compaction caused by wheelings and rolling;
5. Change in land use especially involving the removal of certain crops from steep slopes;
6. Increase organic matter content and structural stability of soil;
7. Work along the contour;
8. Improvement of ditches and field boundaries;
9. Less smooth seed-beds.

At the level of providing day-to-day advice to farmers it is the Agriculture Development and Advisory Service (ADAS) of MAFF, who fulfil this function. One assumes that individual advisers are providing soil conservation advice if and when required and that they are competent to do so. There is, however, no published evidence that training in conservation techniques is ADAS policy or that conservation measures form part of the ADAS experimental programme. It is ironic that the only current ADAS studies of erosion control techniques are funded by the Sugar Beet Research and Education Committee (Davies, 1987).

A further criticism of ADAS policy is that its main thrust is towards short-term maximization of farm yields and profits. The long-term viability of the enterprise is ignored and therefore soil losses over a timescale of tens of years are not taken into account. Farmers are actively encouraged to grow crops such as winter cereals on the South Downs which make short-term economic sense; the long-term consequences of continued erosion on soils that are 15 cm thick are ignored. This conflict between short-term profit and the long-term health of the soils came to the fore in discussions regarding the wisdom of continued growing of cereals at Highdown, Lewes in the light of the high costs due to erosion sustained by householders and local authorities (Stammers and Boardman, 1984). There is clearly a conflict in the roles that ADAS is asked to fulfil, those of providing advice to farmers on both profitability and conservation.

## DOES BRITAIN NEED SOIL CONSERVATION?

The question as to whether Britain needs soil conservation—repeated sixteen years after it was first posed (Evans, 1971)—is answered in the affirmative by data contained in reviews of the erosion problem at a regional or national scale (Boardman and Robinson, 1985; Colborne and Staines, 1985, 1986; Evans and Cook, 1986; Speirs and Frost, 1987; Morgan, 1980, 1985, 1986; Reed, 1979, 1986). However, despite a growing literature on the subject, many people seem unconvinced by the arguments and their objections may be characterized by five assertions each of which requires an answer.

(1) 'The erosion problem is another issue invented by environmentalists etc.' This conspiracy theory has already been referred to. In reply, one has to assess currently available data and note that the more alarming predictions are not from environmental pressure groups who in general remain blissfully unaware of the situation, but from research scientists, e.g. Morgan (1986) suggests that reintroduction of rotations with grass leys is the solution to the problem but, 'if government policy continues to make this solution uneconomic, the alternatives are to accept the consequences of not being able to sustain cereal production on the lighter soils far into the next century'.
(2) 'Erosion is not a national problem but a regional one.' In essence, this assertion is perfectly acceptable with the caveat that, there are as yet large areas of Britain for which no data are available but where, on the basis of soil and land-use characteristics, erosion is likely to be occurring, e.g. in the Cotswolds and on many chalky soils.
(3) 'Erosion is not new: many of our soils have been eroding since they were first farmed by man.' This *laissez-faire* attitude ignores two points: that some soils are now dangerously thin and that modern farming technology and farm management practices can induce erosion on a far greater scale than that of our ancestors. Erosion rates appear to be far higher now and affect much larger proportions of the landscape than formerly.

(4) 'We can do little about erosion since we cannot control the weather.' This assertion is founded in the widespread belief that erosion is primarily a function of heavy rain (e.g. Hudson, 1967). While heavy rain is an important though poorly understood factor (Boardman and Spivey, in press), soil and management factors are often critical: 'erosion is all too often associated with bad management of the land' (Morgan, 1980). Furthermore, it is clear that erosion can occur in the UK under what are elsewhere regarded as light, non-erosive rains (Reed, 1979) and daily totals that are frequently exceeded, e.g. 30 mm in two days on the South Downs with a return period of <1 year.

(5) 'Erosion is a minor cost and small inconvenience to the farmer.' This is undoubtedly true in most cases. However, this may only be true in the short term: the limited data available suggest that the costs will be significant as yields fall on soils that are already thin (Evans, 1981). Alternatively, costs of inputs, particularly fertilizers, will increase as organic matter and the fine fraction of soils are lost by erosion. What is also ignored by such assertions is the off-farm cost of erosion (Evans, 1985; Stammers and Boardman, 1984; Boardman, 1986a; Boardman and Spivey, in press). It is noteworthy that US estimates of off-farm costs suggest that they are far higher than the costs incurred by the farmer (Crosson, 1984). The Conservation Foundation estimates off-farm damage at some $3.1 billion whereas on-farm costs due to loss of crop production are estimated at $40 million per year. The cost of annual soil erosion damage in the state of Ohio is listed in Table 3.2 (from Wall and Dickinson, 1978). Similar evaluation of the total cost of erosion in Britain has not been carried out.

To summarize the arguments in favour of a soil conservation strategy for Britain: erosion is a regional problem and conservation strategies could be directed at high-risk areas; these should be selected with the principle advanced

**Table 3.2** Cost of erosion damages in state of Ohio (from Wall and Dickinson, 1978)

| Activity | Cost/year $ | % of total annual cost |
|---|---|---|
| Sediment dredging from habours on Lake Erie (1976) 3 million cubic metres (>4 million cubic yds) | 7 674 000 | 8 |
| Sediment removal from drains and highway ditches | 23 606 000 | 25 |
| Sediment damage to inland lakes, reservoirs | 4 350 000 | 5 |
| Urban sedimentation | 3 850 000 | 4 |
| Water treatment costs | 8 150 000 | 9 |
| Cropland soil losses | 14 300 000 | 15 |
| Fertilizer losses—N, P and animal waste | 23 356 000 | 25 |
| Energy savings in agriculture | 6 500 000 | 7 |
| Biological values in fish and wildlife | 2 600 000 | 2 |
| Total | 94 386 000 | |

by Larson, Pierce and Dowdy (1983) in mind: 'conservation efforts should be concentrated where erosion damage is greatest, not necessarily where the greatest amount of erosion occurs'. Farming activities are not immediately threatened by erosion but a longer term view suggests that certain areas need protection. In the present climate of commodity over-production conservation makes good sense (Reilly, 1984) although the short-term benefits of conservation measures would be felt mainly by local and water authorities rather than by the farmer.

## PRESENT AGRICULTURAL POLICIES AND CONSERVATION

If by 'public policy' we mean the legislative and financial support from central and local government sources for erosion studies or conservation measures, then little in the way of public policy exists. Indeed, almost all policies work to increase the risk of erosion.

Indirectly many of the measures adopted by farmers in the search for greater efficiency—a success story for British farming—operate in this way. In particular, the decrease in structural stability associated with low organic matter levels (Hodges and Arden-Clarke, 1986); the production of finer seedbeds (Speirs and Frost, 1985); the increase in field sizes and the move to winter cereals all increase the risk of erosion. These developments in farming practice represent aspects of agricultural policy which in the recent past has been geared to the steady increase in crop yields and production of 'cheap' food. With hindsight, we can see that the accounting methods by which 'success' was evaluated were misleading in that they ignored the wider costs of agricultural enterprises such as soil deterioration, ground and surface water contamination, and costs to communities. The wider costs have only become an issue because public awareness of over-production coincided with Third World food shortages. The image of EEC food mountains then became a potent symbol of inadequate, not to say, immoral policies. The relevance of these runaway policies to the erosion debate is clear: the majority of British erosion occurs on winter cereals, a crop the EEC overproduces. In 1986 there were stocks of 24 million tonnes and the cost of the grain mountain to the EEC budget was about £45m per week in storage and disposal costs (Anderson, 1987). In 1984–5 the British taxpayer was paying £100m per year on top of the £358m spent in subsidizing farmers to grow the crops (*Sunday Times*, 1985). The public are in fact subsidizing farmers to erode the land. A national land-use policy which took note of erosion would therefore attempt to:

(1) Reduce the production of winter cereals;
(2) Remove this crop from erodible land;
(3) Educate farmers in conservation practices.

The first of these policies would reduce the food mountain and a decline in erosion would be an incidental gain. It would not necessarily affect the most erodible land and therefore it is desirable to combine the first and second measures. Education is necessary because land-use change can always be reversed and with changing economic circumstances high-risk crops could return to high-risk land: in that case we would have to rely on conservation techniques to control erosion.

## RESEARCH PRIORITIES FOR EROSION AND CONSERVATION

A policy commitment to reduction in erosion and promotion of conservation techniques would involve financial expenditure from public sources on research. At present the main sources of support for government and non-government research activity is funding by the research councils and MAFF; relevant work is also carried out by the Forestry Commission, Institute of Hydrology, the Department of the Environment and the Institute of Terrestrial Ecology.

Six themes for future work on water-borne erosion have been suggested (Boardman, 1987); very similar proposals have been put forward by Davies (1987) and Evans (1987). Most of these would involve collaborative work drawing on funds from several sources. The need for a central organizing group is emphasized in the following section.

(1) The continuation of the SSEW/MAFF air photographic monitoring programme. Some additional areas are needed particularly to assess erosion on chalky soils in southern and eastern England and also the Cotswolds. Some areas in which in the first five years of monitoring low numbers of fields have been affected, should be discontinued. The project is also important because there is always the possibility that land use change will bring into arable use soils that under grazing are not erodible.
(2) Processing and evaluation of the data from the monitoring programme should be given high priority. Evaluation will need to take into account current land use and future changes: for example the intensification of agriculture on erodible soils. Climatic data must also be considered.
(3) Erosion risk maps should be prepared, initially for selected high-risk areas.
(4) Little data are available in the UK for off-farm effects of erosion: much exist but need to be collated. Effects include the costs of road clearance, costs of drainage disruption and pollution of watercourses and reservoir siltation. Current work on the Wiltshire Avon suggests that runoff from eroding fields affects fisheries.
(5) There is little work on conservation techniques and most farmers are unable or unwilling to attempt to control erosion. This is both a problem for the development of suitable low-cost techniques and the communication of this information to the farmer.

(6) The ability to predict erosion rates for given slope, soil, land use and rainfall conditions is needed. Computer models require adequate databases which both field mapping and air photographic surveys are providing. Initial testing of suitable models has been carried out on very limited data bases using mainly non-British data (e.g. Morgan, Morgan and Finney, 1984). Work using expert systems is yielding promising results and could be extended (Harris and Boardman, 1988).

## A CONSERVATION POLICY: THE WAY FORWARD

Prior to an assessment of the prospects for Britain it is worth noting that Canada, a country not altogether dissimilar in terms of land use and climate, has recently become aware of the threat posed by erosion. Although researchers have for some time reported instances of erosion and discussed the causes, two recent developments have brought the issues into the arena of public policy. The first of these is a concern with water quality especially as regards water entering the Great Lakes system. This led to the setting up of the Pollution from Land Use Activities Reference Group (PLUARG) which has sponsored many studies particularly addressing the problem of phosphorus transport in runoff from arable fields (e.g. Wall, Dickinson and Van Vliet, 1982). The realization that water quality is intimately linked to erosion of agricultural land has focused attention on the need for soil conservation. The second development has been a more general concern with the degradation of Canadian soils. This is reflected in the Sparrow Report (1984). The preface to this report from the Standing Committee on Agriculture, Fisheries and Forestry, by Senator Sparrow, notes that by increasing the awareness of soil degradation, 'the Committee hopes to make soil conservation a national issue'. The Sparrow Report was followed by an equally impressive document from the Science Council of Canada (1986). In its recommendation the emphasis is on increased funding for conservation research and the efficient transfer of knowledge to the farmer. A co-ordinated national conservation programme is recommended. For the British reader these two documents make salutary reading in view of the complacent attitude of MAFF to similar problems. However, the dramatic increase in Canadian awareness and associated action at a government level (e.g. PLUARG), shows that a country need not sit back and wait until soils become uneconomic to farm or off-farm costs force changes upon the agricultural sector.

What are the essential elements of a conservation policy for Britain? First, one must recognize that the transition from doing nothing to doing what is necessary is likely to be difficult in that little money will be available for direct funding of soil conservation. However, money is now available which can be used to reduce erosion on agricultural land. The setting up by MAFF of the Environmentally Sensitive Area (ESA) scheme allows farmers in six designated areas (to be expanded to twelve) to apply for grant aid in order to retain traditional agricultural landscapes. On the eastern South Downs, for example, two levels of grant are available for retention of unimproved grassland and reversion of arable fields to grass. In terms of soil conservation it is the latter grant that is relevant, allowing farmers to take high-risk winter cereals out of erodible fields; one of the claims made for reversion to grassland being that it, 'promotes soil stabilization on steep slopes liable to soil erosion' (MAFF, 1986). Unfortunately at present the scheme is entirely voluntary, it is unmonitored by any outside agency due to confidentiality clauses in the management agreements, and farmers are critical of the level of payments. There is as yet no evidence that fields which are regularly eroding are being taken out of cereals. For this potentially useful scheme to contribute to reductions in erosion some form of targeting of grant monies on high-risk sites is desirable. Such sites can be identified (Boardman, 1986b) and therefore a priority system for the allocation of grants could be established.

A simple method for identification of high-risk sites is illustrated in Table 3.3. Sites of major erosion are selected from a data base containing about 250 sites from a sample area of the South Downs of 36 $km^2$. Sites at risk are characterized by a slope length of >200 m, a maximum angle of >10°, relief within the field of >30 m, they grow winter cereals drilled in a downslope direction and have a rainfall factor for the growing season >5. The rainfall factor is calculated by allocation of a value of 1 to events of >30 mm in 2 days and a value of 2 to events of >30 mm in 1 day. Sites with these physical characteristics may be identified on maps. An additional category of high-risk sites is where runoff is likely to enter properties or roads or affect drainage systems. A more sophisticated analysis of the databases for 1982–7 has now been undertaken using an expert system approach. This is potentially more useful and flexible as a conservation tool than the previous approach involving inspection of a data base (Harris and Boardman, 1988). Both approaches are dependent on provision of an accurate data base built up by geomorphological mapping and measurement of eroding sites over several years. For the ESA scheme to play a role in the control of erosion targeting of grant money is essential. This would ensure that public money is not wasted on sites of low risk. Cost–benefit analysis would undoubtedly show that targeting is justified with grant monies from MAFF benefiting local authorities who suffer the principal short-term losses as a result of erosion.

The ESA scheme, as implemented on the South Downs, is a form of set-aside which has been proposed as a conservation measure for the UK (Burnham, 1985; Potter, 1987). Set-aside schemes address the problem of over-production and the need to take out of productive agricultural use about 800 000 hectares in the next few years. These schemes also recognize that the central problem of instituting conservation measures is a financial one. Farmers tend to grow the most profitable crops that their land will sustain; this may result in erosion.

**Table 3.3** Eastern South Downs: sites of major soil erosion, 1982–5

| Site | Area (ha) | Slope length (m) | Max. angle (deg.) | Type | Relief (m) | Rate ($m^3$/ha) | Rainfall factor | Cultivated |
|---|---|---|---|---|---|---|---|---|
| 8202 | 15 | 400 | 10 | vs | 63 | 7 | 11 | ds/win/roll |
| 8203 | 19 | 400 | 11 | vs | 59 | 32 | 11 | ds/win/roll |
| 8211 | 7 | 450 | | vb | 50 | 7 | 9 | ac/win |
| 8222 | 5 | 300 | 11 | vb and vs | 38 | 4 | 11 | ds/win |
| 8240 | 36 | 1710 | | vb | 92 | 3 | 11 | ds/win |
| 8257 | 17 | 580 | 15 | vb and vs | 53 | 6 | 6 | ds/win |
| 8261 | 14 | 440 | 22 | vb and vs | 42 | 6 | 11 | ds/win |
| 8267 | 17 | 590 | 22 | vb and vs | 49 | 6 | 11 | ds/win |
| 8268 | 18 | 990 | 22 | vb and vs | 69 | 6 | 11 | ds/win |
| 82.D | 4 | 200 | 12 | vs | 38 | 38 | 5 | ds/win |
| 8407 | 8 | 450 | 18 | vs | 85 | 10 | 6 | ds/plo |
| 84.B | 0.5 | 130 | 11 | vs | 11 | 10 | 3 | ds/plo |
| 8505 | 2 | 260 | | vb | 38 | 5 | 7 | ds/win |
| 8511 | 6 | 540 | | vb | 60 | 12 | 7 | ds/win |
| 8518 | 7 | 180 | 15 | vs | 45 | 14 | 7 | ds/win |
| 8519 | 2 | 240 | | vb | 17 | 4 | 7 | ds/win |
| 8520 | 3 | 300 | 11 | vb and vs | 38 | 6 | 7 | ds/win |
| 8521 | 8 | 450 | 18 | vs | 85 | 13 | 7 | ds/win |
| 8525 | 2 | 110 | 15 | vs | 16 | 5 | 7 | ds/win/roll |
| 8526 | 8 | 420 | | vb | 46 | 4 | 7 | ds/win |
| 8536 | 1 | 120 | 16 | vs | 22 | 8 | 7 | ds/win |
| 85.C | 2 | 400 | 9 | vs | 57 | 30 | 7 | ds/win |
| Typical site: | | >200 | >10 | | >30 | | >5 | ds/win |

vs = valley side; vb = valley bottom; ds = downslope; ac = along contour; win = winter cereal; plo = plough; roll = rolled.
Maximum angle in field is not recorded for cases of valley bottom erosion.

In order to persuade them to do otherwise (for the public good) financial incentives are necessary.

Set-aside schemes may not, however be an unmitigated success in terms of reduction of erosion. One scenario is that with continuing increases in yields all British cereals could be grown in East Anglia on highly efficient, specialized farms. This would undoubtedly lead to a sharp increase in erosion rates in that area. Whether, on a national scale, this would be offset by decreases elsewhere, e.g. the South Downs, is not known. But it is not a system that could be continued into the future without attempts to control erosion in the high yielding, grain growing areas.

Advice to farmers on soil conservation techniques should be available and not simply as a response to severe erosion events. Further publications detailing appropriate procedures should be produced (cf. Evans, 1984); preferably, these would be widely disseminated and free of charge.

One organization should be given responsibility and funds in order to carry out monitoring, research and the provision of advice. Historically, the first function has been carried out by the Soil Survey, the second by no public body and the third, in so far as any advice was given, by ADAS. This could continue although it is not a satisfactory arrangement, not least because of the Soil Survey's loss of all personnel who were actively engaged in erosion work and ADAS's traditional obsession with productivity. What has been called 'the greening of ADAS', and their move into the sphere of providing conservation advice, offers hope for the future. At present a confused situation exists as regards responsibility for reducing erosion and indeed for providing the basic data for evaluation of the problem, all the latter having come from the Soil Survey and academic researchers. Because of this confusion there is no clear route by which information and funds can flow. The ideal solution would be a Soil Conservation Unit within MAFF with a clear responsibility to collect data, advise farmers and liaise with non-government researchers. Liaison would also be necessary with local government, water authorities and others affected by erosion, perhaps via the Department of the Environment. None of these responsibilities would be difficult to fulfil: data collection is already well established; liaison between various concerned bodies does occur in times of crisis and, in terms of advice to farmers, we have fifty years of North American experience to draw upon and adapt.

What are the prospects for the implementation of such a modest scheme or indeed any concerted effort to reduce erosion? If we assume that in the next few years, contrary to all logic, land-use policy will not radically change the relationship between high-risk crops and high-risk land, then progress is probably dependent on some mild form of catastrophe, for instance, a winter on the South Downs with 25 per cent more rainfall than 1982. This would produce significantly more erosion than in 1982. In the interim, what has changed is public and media interest in the environment and the ability of researchers to direct media attention to these issues. There has also been a growth in awareness on the part of local and water authorities that soil on roads and in houses, and the pollution of rivers, is a result of mismanagement by man and not divine intervention. Will Lewes District Council or West Derbyshire District Council be prepared to spend yet again £12 000 or £20 000 respectively to protect householders?

Recent events on the South Downs have again directed attention to this issue. On 7 October 1987 runoff from winter cereal fields at Rottingdean caused an estimated £1m worth of damage to 80 houses. Similar floods carrying soil from cereal fields affected housing in seven other areas of the Brighton conurbation, some of which had been affected in 1976 and 1982. Emergency protective work by Brighton Borough Council will cost in excess of £100 000 and the prospect of legal action by householders to recover uninsured losses remains. These cases are replicas of the flooding and damage to the Highdown estate in Lewes in 1982 but on a far larger scale. At that time proposals were put forward to prevent a repetition and these include a code of practice to be prepared by MAFF for farmers in high-risk situations; nothing was done (Boardman, Stammers and Chestney, 1983; Stammers and Boardman, 1984).

This rather pessimistic appraisal of how change may come about is based on the belief that changed attitudes within MAFF will only occur if political pressure is brought to bear, e.g. from local authorities or MPs. In many areas of environmental concern in the last few years significant changes of policy have occurred due to a coincidence of interests of usually antagonistic groups. In this case the growth of environmental awareness coincides with the problem of over-production and the need to reduce the spending on agricultural subsidies. In this context soil conservation measures can be regarded as part of a package of reforms which would fulfil the demands of apparently conflicting interest groups, those bearing the banners of environmental conservation and government expenditure. They would also be in line with the more radical approach to conservation issues of many of our European Economic Community partners.

## Acknowledgement

I would like to thank Dr R. Evans for permission to reproduce Figure 3.1 and for valuable comments on a first draft of the text.

## REFERENCES

Anderson, J. A. (1987). Policy options. In D. J. L. Harding (ed.) *Agricultural Surpluses? Environmental Implications of Changes in Farming Policy and Practice in the U.K.*, Proceedings Symposium, Institute of Biology, London.

Boardman, J. (1983a). Soil erosion at Albourne, West Sussex, England, *Applied Geography*, **3**, 317–29.
Boardman, J. (1983b). Soil erosion on the Lower Greensand near Hascombe, Surrey, 1982–3, *Journal Farnham Geological Society*, **1**(3), 2–8.
Boardman, J. (1984). Erosion on the South Downs, *Soil and Water*, **12**(1), 19–21.
Boardman, J. (1986a). The context of soil erosion, *SEESOIL*, **3**, 2–13.
Boardman, J. (1986b). *Soil Erosion in a Proposed Environmentally Sensitive Area*, Report for East Sussex County Council.
Boardman, J. (1987). *Current Work and Proposals for Collaborative Research Work in the Future*, unpublished paper for Review Meeting on soil erosion, MAFF, January 1987, London.
Boardman, J and J. Hazelden (1986). Examples of erosion on brickearth soils in east Kent, *Soil Use and Management*, **2**(3), 105–8.
Boardman, J. and D. A. Robinson (1985). Soil erosion, climatic vagary and agricultural change on the Downs around Lewes and Brighton, autumn 1982, *Applied Geography*, **5**, 243–58.
Boardman, J. and D. Spivey (in press). Flooding and erosion in west Derbyshire, April 1983, *East Midlands Geographer*.
Boardman, J., R. L. Stammers and D. Chestney (1983). *Flooding Problems at Highdown, Lewes: Technical Report*, Report to Lewes District Council.
Burnham, P. (1985). Curing the grain pain, *The Countryman*, Summer 1985, 137–41.
Clark, M. J. (1976). Barton does not rule the waves, *Geographical Magazine*, July 1976, 581–5.
Colborne, G. J. N. and S. J. Staines (1985). Soil erosion in south Somerset, *Journal Agricultural Science, Cambridge*, **104**, 107–12.
Colborne, G. J. N. and S. J. Staines (1986). Soil erosion in Somerset and Devon, *SEESOIL*, **3**, 62–71.
Crosson, P. (1984). New perspectives on soil conservation policy, *Journal Soil and Water Conservation*, **39**, 222–5.
Davies, D. B. (1987). *Soil Erosion in England and Wales*, unpublished paper for Review Meeting on soil erosion, MAFF, January, 1987. London.
Davies, D. B., D. J. Eagle and J. B. Finney (1972). *Soil Management*, Farming Press, Ipswich.
Davies, D. B., D. J. Eagle and J. B. Finney (1982). *Soil Management*, Fourth edn, Farming Press, Ipswich.
ENDS (1984). Environmental Resources 23—Soil Erosion: an unsustainable face of modern farming. In *ENDS Report* 115, pp. 9–10, Environmental Data Services Ltd, London.
Evans, R. (1971). The need for soil conservation, *Area*, **3**(1), 20–23.
Evans, R. (1981). Assessments of soil erosion and peat wastage for parts of East Anglia, England. A field visit. In R. P. C. Morgan (ed.) *Soil Conservation: Problems and Prospects*, pp. 521–30, Wiley, Chichester.
Evans, R. (1984). *Soil Erosion by Water*, ADAS, MAFF, Leaflet 890.
Evans, R. (1985). *Soil Erosion—the Disappearing Trick*, unpublished paper to conference on Better soil management for cereals and oilseed rape, Royal Agricultural Society, England.
Evans, R. (1987). *Erosion What Needs to be Done*, unpublished paper for Review Meeting on soil erosion, MAFF, January 1987, London.
Evans, R. and S. Cook (1986). Soil erosion in Britain, *SEESOIL*, **3**, 28–58.
Evans, R. and S. Nortcliff (1978). Soil erosion in north Norfolk, *Journal Agricultural Science, Cambridge*, **90**, 185–92.

Evans, R. and D. Skinner (1987). A survey of water erosion, *Soil and Water*, **15**, 28–31.
Frost, C. A. and R. B. Speirs (1984). Water erosion of soils in south-east Scotland—a case study, *Research and Development of Agriculture*, **1**, 145–52.
Fullen, M. A. (1985). Erosion of arable soils in Britain, *International Journal of Environmental Studies*, **26**, 55–69.
Harris, T. M. and J. Boardman (1988). *Expert System Approaches to Soil Erosion Risk Assessment*, unpublished paper for IBG/BGRG Symposium, January 1988.
Harris, T., R. Evans and J. Boardman (in preparation). The use of aggregate statistics for water borne erosion data: some cautionary comments.
Hodges, R. D. and C. Arden-Clarke (1986). *Soil erosion in Britain: Levels of Soil Damage and their Relationship to Farming Practice*, The Soil Association, Bristol.
Hudson, N. W. (1967). Why don't we have soil erosion in England? In J. A. C. Gibb (ed.) *Proceedings of Agricultural Engineering Symposium*, Institute Agricultural Engineers Paper 5/B/42.
Larson, W. E., F. J. Pierce and R. H. Dowdy (1983). The threat of soil erosion to long-term crop production, *Science*, **219**, 458–65.
Mackney, D., J. M. Hodgson, J. M. Hollis and S. J. Staines (1983). *Legend for the 1:250,000 Soil Map of England and Wales*, Soil Survey of England and Wales, Harpenden.
MAFF (1986). *Guidelines for Farmers, SD/ESA/4, South Downs Environmentally Sensitive Area*, Ministry of Agriculture, Fisheries and Food.
Mather, A. S. (1983). Land deterioration in upland Britain, *Progress in Physical Geography*, **7**, 210–28.
Morgan, R. P. C. (1980). Soil erosion and conservation in Britain, *Progress in Physical Geography*, **4**, 24–47.
Morgan, R. P. C. (1985). Assessment of soil erosion risk in England and Wales, *Soil Use and Management*, **1**, 127–30.
Morgan, R. P. C. (1986). Soil erosion in Britain: the loss of a resource, *The Ecologist*, **16**, 40–41.
Morgan, R. P. C., D. D. V. Morgan and H. J. Finney (1984). A predictive model for the assessment of soil erosion risk, *Journal Agricultural Engineering Research*, **30**, 245–53.
Phillips, J., D. W. Yalden and J. H. Tallis (1981). *Moorland Erosion Study. Phase 1 Report*, Peak Park Joint Planning Board, Bakewell.
Pidgeon, J. D. and J. M. Ragg (1979). Soil, climatic and management options for direct drilling cereals in Scotland, *Outlook on Agriculture*, **10**, 49–55.
Potter, C. (1987). The conservation alternative. In D. J. L. Harding (ed.) *Agricultural Surpluses? Environmental Implications of Changes in Farming Policy and Practice in the U.K.*, Proceedings Symposium, Institute of Biology, London.
Reed, A. H. (1979). Accelerated erosion of arable soils in the United Kingdom by rainfall and run-off, *Outlook on Agriculture*, **10**, 41–8.
Reed, A. H. (1983). The erosion risk of compaction, *Soil and Water*, **11**, 29–33.
Reed, A. H. (1986). Soil loss from tractor wheelings, *Soil and Water*, **14**, 12–14.
Reilly, W. K. (1984). Soils, society and sustainability, *Journal Soil and Water Conservation*, **39**, 286–90.
Science Council of Canada (1986). *A Growing Concern: Soil Degradation in Canada*, Science Council of Canada, Ottawa.
Soil Association (1986). *Soilwatch*, Soil Association, Bristol.
Sparrow Report (1984). *Soils at Risk, Canada's Eroding Future*, Standing Senate Committee on Agriculture, Fisheries and Forestry, Senate of Canada, Ottawa.

Speirs, R. B. and C. A. Frost (1985). The increasing incidence of accelerated soil water erosion on arable land in the east of Scotland, *Research and Development in Agriculture*, **2**, 161–7.

Speirs, R. B. and C. A. Frost (1987). Soil water erosion on arable land in the United Kingdom, *Research and Development in Agriculture*, **4**, 1–11.

Spoor, G. (1986). Machinery matters, *Crops*, February 2, 1986.

Stammers, R. L. and Boardman, J. (1984). Soil erosion and flooding on downland areas, *The Surveyor*, **164**, 8–11.

*Sunday Times* (1985). UK pays dear for grain glut, *Sunday Times*, 5 May 1985.

Tallis, J. H. and D. W. Yalden (1983). *Peak District Moorland Restoration Project. Phase 2 Report: Re-vegetation Trials*, Peak Park Joint Planning Board, Bakewell.

Wall, G. J. and Dickinson, W. T. (1978). Economic impact of erosion, In *Soil Erosion in Ontario, Notes on Agriculture*, **14**, 3, 10–12.

Wall, G. J., W. T. Dickinson and L. J. P. Van Vliet (1982). Agriculture and water quality in the Canadian Great Lakes Basin: II Fluvial sediments, *Journal Environmental Quality*, **11**, 482–6.

Geomorphology in Environmental Planning
Edited by J. M. Hooke



# 4 Soil Erosion Control: Importance of Geomorphological Information

**ROY MORGAN and JANE RICKSON**
*Silsoe College, Silsoe, Bedford*

## INTRODUCTION

The consequences of soil erosion, such as declining soil productivity on-site, pollution and sedimentation off-site and the unsightly scarring of the landscape, are well known. They should be of concern to all those involved in the management of our soil and water resources. Although climate, soil, landform and vegetation cover all interact to determine how much erosion takes place, it is the vegetation which ultimately controls how much protection is afforded to the land. Environmental and agricultural policies which bring about changes in land use therefore affect the severity of erosion. This means that policy-makers and planners whose decisions influence the way in which our land is used are directly responsible for how well erosion is controlled.

Since geomorphologists study the form and materials of the land surface and the processes operating thereon, they are able to understand and predict the consequences of changes in land use. They can therefore help the policy-maker in predicting the likely effects of different policies and in developing strategies to mitigate these effects.

Policies for soil erosion control need to be based on knowledge of the severity of erosion and the nature of the processes involved. Vital background information for soil conservation planning comes from an evaluation of the resources of the land with respect to its suitability for different uses and from research into the mechanics of water and wind erosion, all of which the geomorphologist is well-equipped to provide. Indeed, there are dangers in not seeking the advice of the geomorphologist in designing systems for erosion control because they are then often based on inadequate knowledge. This danger is illustrated in this chapter by examining the role of vegetation and geotextiles in controlling erosion. Attention is drawn to the need to review how these materials are used and the need to question a number of long-held beliefs on how they influence erosion.

## VEGETATION AND EROSION CONTROL

Vegetation is the basis for agronomic or biological methods of erosion control. Erosion on arable land is minimized using vegetation as cover crops during fallow periods in a crop rotation system, as ground cover under tree crops and as barrier strips placed on the contour to intercept runoff or at right angles to wind. Controlled and rotational grazing helps to maintain a dense vegetation cover and reduce erosion on rangeland. Vegetation is also being increasingly used by civil engineers for slope stabilization and surface erosion control.

Numerous studies demonstrate the effectiveness of a vegetation cover in reducing erosion compared with a bare soil. However, the relationship between soil loss and percentage vegetation cover is not clear cut. As Figure 4.1 shows, a reduction in erosion to 50 per cent of that from bare ground can be achieved by a 10, 40 or 70 per cent vegetation cover depending on which relationship is chosen. Clearly this level of understanding is inadequate for design. Vegetation cannot yet be considered as an engineering material described by a set of design parameters which can be used to select appropriate vegetation types to treat specific erosion problems. Instead, the effects of vegetation are described by a single number or coefficient, such as the C-factor in the Universal Soil Loss Equation. This can be misleading because when the vegetation effects are broken down into their components of interception of rainfall, plant-induced roughness

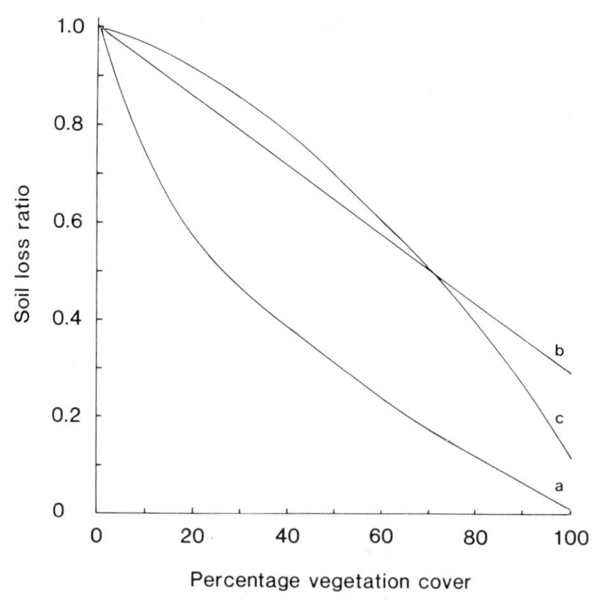


**Figure 4.1** Examples of relationships between soil erosion and percentage vegetation cover: (a) ground level vegetation (Wischmeier, 1975); (b) canopy level at 1 m above the ground (Wischmeier, 1975); and (c) oat straw mulch (Singer and Blackard, 1978)

on flow and root reinforcement of the soil, it is found that vegetation is not always beneficial and under some circumstances is capable of enhancing rather than inhibiting erosion.

The present state of understanding of the effect of vegetation on soil aggregate breakdown and detachment by raindrop impact is that detachment decreases exponentially with increasing cover if that cover is on the ground surface. Experimental evidence to support this comes from studies with crop residues (Laflen and Colvin, 1981), grass (Lang and McCaffrey, 1984) and stones

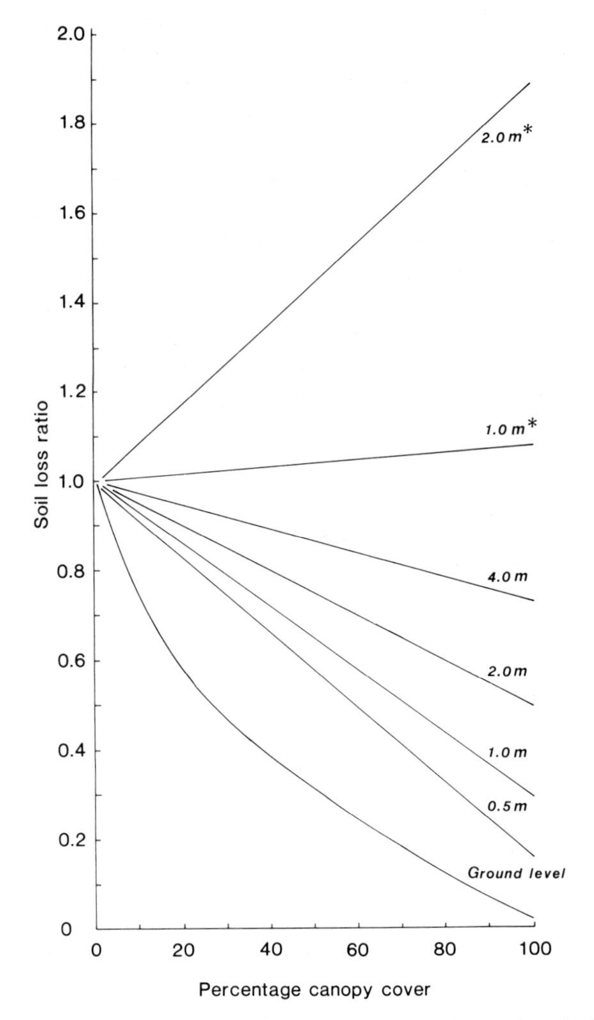


**Figure 4.2** Relationships between soil loss ratio for soil detachment by raindrop impact and percentage vegetation cover for different canopy heights. Asterisks refer to canopies which generate large diameter leaf drips. After Wischmeier (1975) and Styczen and Høgh-Schmidt (1986)

(van Asch, 1980). When the canopy is above ground two possible relationships apply. Generally, detachment decreases linearly with increasing cover (Wischmeier, 1975; Foster, 1982) but, under some circumstances, for example if large quantities of leaf drip are generated on the canopy, detachment may increase linearly (Styczen and Høgh-Schmidt, 1986) (Figure 4.2). Experimental support for the latter relationship comes from field studies of soil detachment under maize (Morgan, 1985) where, with 88 per cent canopy cover, detachment rates were one-and-a-half times those on bare soil despite a reduction in rainfall volume at the ground to 40 per cent.

The effectiveness of rain beneath a plant cover in detaching soil particles is demonstrated by rainfall simulation studies with Brussels sprouts and with toy oak trees (Morgan *et al.*, 1986) which reveal a pattern of decreasing soil detachment with increasing percentage canopy up to 15–20 per cent and thereafter an increase so that by the time 50–60 per cent cover is attained soil detachment rates are similar to those for bare soil. In this last case, the effects of increasing rainfall interception and therefore decreasing volume of rainfall at the ground surface as the plant grows are offset by changes in the drop-size distribution of the rain due to formation of leaf drip on the canopy.

Large leaf drips, typically 5 to 6 mm in diameter, falling from 1 m, are more efficient agents of detachment than drops of 2 to 3 mm at terminal velocity in natural rainfall (Finney, 1984; Table 4.1). Research is needed to classify plants according to their ability to form leaf drip. The size and frequency of leaf drip is related to the length-area ratio of the leaf. Long narrow leaves channel water rapidly to their tips resulting in frequent but relatively small (3.5 mm) drops from a single point whereas leaves with ratios less than 0.015, typical of mature sugar-beet and brassicas, produce less frequent but larger (6.0 mm) drops from a small number of points. These large drops are capable of greater soil

**Table 4.1** Soil detachment by single raindrops

| Drop diameter (mm) | Detachment of sand (0.15–0.21 mm)* (g/J) | Detachment of sandy loam soil[†] (g/J) |
|---|---|---|
| | *Simulation of natural rainfall* | |
| | *(infinite fall height)* | |
| 2 | 16 | 13 |
| | *Simulation of leaf drip* | |
| | *(1 m fall height)* | *(10–65 cm fall height)* |
| 3 | 10–14 | |
| 5 | 16–17 | |
| 6 | 25–31 | 1–40 |

*Data from Matthew Stuttard and Jane Brandt (personal communication).
[†]From Finney (1984).

**Table 4.2** Leaf drip generation

| Leaf length/ area ratio | Number of drip points | Drip frequency (s) | Drop diameter (mm) |
|---|---|---|---|
| 0.007 | 15 | 12.0 | 5.8 |
| 0.013 | 6 | 15.8 | 6.2 |
| 0.017 | 4 | 9.3 | 6.4 |
| 0.026 | 1 | 0.6 | 3.5 |

Data from Carole Andrews (personal communication).

detachment (Andrews, 1986; Table 4.2). Little is known about other factors that may influence leaf drip, such as the angle of inclination of the leaf and the roughness of the leaf surface. Noble and Morgan (1983), however, found that although the angle of inclination of the leaf petiole to the main stem made a minor but significant contribution to the volume of leaf drip under a Brussels sprout plant, it did not affect the rate of soil detachment.

There has been less questioning of the generally accepted relationships that runoff volume decreases linearly and runoff velocity exponentially with increasing cover (Rickson and Morgan, 1988) and that soil erodibility increases exponentially with decreasing root density (Dissmeyer and Foster, 1985). This may be because fewer confirmatory studies have been made by geomorphologists. Further investigations are certainly justified by the findings of De Ploey, Savat and Moeyersons (1976) who recorded increasing erosion by scour due to concentration of flow between plant stems, and Yair and Lavee (1976) who observed a similar effect due to flow concentration between stones. Flow concentrations can also arise at the base of plant stems as a result of stemflow and localized soil erosion can occur (Herwitz, 1986).

Further evidence that flow concentration between plant stems can enhance the likelihood of erosion comes from studies of wind velocity close to the ground surface around single crop rows (Morgan and Finney, 1987). Normally, air is slowed as a result of drag imparted to the flow as it passes through a crop row. This effect becomes less pronounced in stronger winds so that the drag coefficient of a crop decreases with increasing wind speed. However, under conditions of moderate and continuous rather than fluctuating wind speeds and early stages of crop growth, the drag coefficient for plants in the lower 5 cm of the atmosphere increases with wind velocity. This is attributed to a wall effect in the atmosphere created by the continuous movement of the leaves. Wind velocity is increased over the top of the wall and through gaps in the wall between the plant stems. This results in localized increases in shear velocity and therefore in greater risk of sediment movement.

Clearly the effects of vegetation on erosion are more complex than have been generally appreciated. Geomorphologists can help to unravel these effects by studies of how vegetation behaves in affecting detachment and transport of soil particles by wind and water. Attention must be given to identifying the key properties of the vegetation such as height, plant morphology, leaf shape and root structure which affect how well the soil is protected. These studies will provide the base for a classification of plants according to their effectiveness in erosion control. This will allow the development of a design procedure for soil conservation in which the requirements of a vegetation cover are specified for a given level of tolerable erosion and different vegetation types are evaluated for their ability to meet these requirements. The geomorphologist will then be able to advise the policy-maker on how and what vegetation types should be used to control erosion and on what the risk is of growing specific crops or plants at particular sites.

## GEOTEXTILES

Where vegetation is difficult to establish because of a hostile soil and climatic environment, as on road embankments and cut slopes, erosion problems are being solved by the use of geotextiles. A geotextile is any permeable textile material used with foundation, soil, rock, earth or any geotechnical engineering-related material (Giroud and Caroll, 1983). It can be in the form of a mat, sheet, grid or web of natural or artificial fibre which is either placed on the slope or buried in the top soil. Unfortunately, theoretical investigations of how these products work have lagged behind their use so that few engineers are fully conversant with the fundamentals of geotextiles (Anon., 1986). This can result in incorrect use and application of the various geotextiles currently available and in greater costs being incurred through failure. The gap in fundamental knowledge can be satisfied by geomorphological research. Research is needed to formulate design specifications for selection and use of the materials. Design guidelines are urgently needed by practising civil engineers to ensure optimum use and avoidance of misuse of the products in a variety of sites and environments. Such guidelines would complement the existing use of geomorphologically based design parameters in highway construction (Brunsden, Doornkamp and Jones, 1975).

When evaluating the problem of erosion at a specific site, such as a road embankment, the following questions need to be answered before a solution either with or without geotextiles can be proposed.

(1) How severe is the erosion problem? To what extent is the top soil removed so that vegetation, regarded potentially as the best controller of erosion (Hudson, 1981), cannot establish itself? If this is the case we have identified the need to apply a geotextile to the site.
(2) Is the erosion natural or accelerated by man, either intentionally or otherwise?

(3) How fast is erosion occurring and where does the sediment go? Does it silt up drainage ditches or roadside ditches? If so, what are the off-site costs of erosion and the implications of these? Would the use of geotextiles reduce these costs?
(4) What are the erosion processes operating at the site? This is important as different geotextiles act in different ways. Some only reduce soil detachment by rainfall without changing the transporting capacity of any generated runoff. Others, through their inherent properties, reduce runoff transport capacity without protecting the soil from raindrop impact.

Answers to these fundamental questions can be found by taking a geomorphological approach to site assessment. The geomorphologist is trained to appreciate which specific erosion processes are operating and at what rate. This knowledge is vital in evaluating the consequences of erosion both on- and off-site.

Having identified the nature of the erosion problem and selected geotextiles as a solution, the question arises of which product to use. The answer is not yet easily determined but can be found with a geomorphological approach involving detailed research to assess how different geotextiles affect individual erosion processes. Geotextiles may be expected to perform two functions in relation to soil erosion control: (1) they provide temporary or permanent control of erosion by directly affecting the erosion processes; and (2) they provide a stable environment for vegetation establishment, development and regeneration.

In broad terms, geotextiles work by mimicking the salient properties of vegetation that help to reduce erosion. These are:

(1) Plant cover: provided by surface erosion mats which ideally cover 65 per cent or more of the soil surface and protect it from raindrop impact; since the cover is on the ground, plant canopy effects of concentration of rainfall and formation of leaf drip do not arise;
(2) Plant stems: simulated by the mesh of surface erosion mats which impart roughness to and reduce the velocity of flow;
(3) Root mat: simulated by the cell and mesh structures of buried erosion mats which effectively reinforce the soil, creating an increase in cohesion and shear strength, and enhance infiltration of surface water;
(4) Organic matter: provided by biodegradable and light-degradable mats.

Thus the relationships between vegetation and erosion are copied by the use of geotextiles with the advantage that geotextiles need no time to establish or develop these salient properties that reduce erosion.

At present there is limited research to show that geotextiles are effective in reducing erosion (Table 4.3) but no studies have been made to demonstrate how the products work. It is possible to test which geotextiles are best in reducing

**Table 4.3** Sediment yield reduction with geotextiles

| Treatment | Sediment yield (g) | Sediment yield as % of that from unprotected soil |
|---|---|---|
| Control | 263 | 100 |
| Surface erosion mats | 57 | 22 |
| | 84 | 32 |
| Buried erosion mats | 189 | 72 |
| | 124 | 47 |
| Cellular geotextile | 136 | 52 |

soil detachment by rainsplash using standard splash erosion experiments with splash cups where no overland flow is generated, and a rainfall simulator. The extent to which splash is reduced by each geotextile relative to a control target cup can be quantified for a number of soils and for storms representative of chosen frequencies, durations and magnitudes that may occur at any given site. Similarly, geotextiles can be tested for their ability to reduce runoff volume and velocity using a runoff rig (Quansah, 1985) designed to have variable slope, soil type and discharge. The geotextiles can be tested for their properties of roughness, depression storage and moisture absorption that help to reduce the transport capacity of the runoff. Research is also needed to examine the effect of geotextiles on soil strength; previous work by engineers has concentrated on the strength of the geotextile material itself but not on its effect when combined with soil (Masounave, Denis and Rollin, 1980). From these experiments geomorphologists can obtain a set of data that can be used to define design parameters for application of geotextiles for soil erosion control.

Another research need is to assess the relationship of geotextiles to eventual vegetation establishment and growth. Vegetation as a long-term controller of erosion is to be preferred because (a) it is self-regenerating, (b) involves minimal maintenance costs, (c) is environmentally acceptable, (d) is aesthetically pleasing and (e) has inherent engineering properties. At present there is no work on the effect of geotextiles on the soil/plant environment, e.g. microclimate, soil moisture status, stability of the soil for vegetation establishment, protection against wind erosion, enhancement of organic matter and nutrients by biodegradable products. Geomorphologists are able to assess and evaluate these effects as a result of their backgrounds in biogeography, soil science and ecology.

Related to this interrelationship between geotextiles and vegetation is the question of whether the vegetation alone, when fully established, can stabilize a slope and reduce erosion or only when in combination with a geotextile. This would affect the choice between a biodegradable or permanent geotextile. Also, does the combination of a geotextile and vegetation allow greater angles of safety when used on embankments and cuttings? Will this affect land requirements for road construction?

The environmental impact of geotextiles, an area hardly touched on up to now, can also be assessed by geomorphologists. For example, what are the implications of using a geotextile of artificial fibre? Are geotextiles more aesthetically pleasing than concrete yet provide the same slope stabilization benefits? What are the ecological effects of permanent erosion mats on a slope? Answers to these questions are vital for the efficient and effective use of geotextiles in soil erosion control.

## CONCLUSIONS

Geomorphological research can aid decision-making on the use of vegetation and geotextiles for erosion control. At present their use is hampered by the lack of clear-cut design procedures based on identification of the erosion process at a site, specification of the requirements of vegetation and geotextiles for controlling erosion at that site, and the matching of possible plant communities and geotextile products to those requirements according to easily definable parameters. There is an urgent need to provide agricultural engineers and civil engineers with technical guidance on the design properties of vegetation and geotextiles. The geomorphologist is well-equipped to fill this need.

## REFERENCES

Andrews, C. (1986). The effect of leaf size and shape on the production of leaf drip and soil detachment under plant covers, Unpub. MSc Thesis, Silsoe College.

Anon. (1986). Frontispiece to *Geotextiles and Geomembranes Journal*, (Ed. T. S. Ingold), Elsevier.

Brunsden, D., J. C. Doornkamp and D. K. C. Jones (1975). Applied geomorphology: a British view. In C. Embleton, D. Brunsden and D. K. C. Jones (eds), *Geomorphology: Present Problems and Future Prospects*, pp. 251–62, Oxford Univ. Press.

De Ploey, J., J. Savat and J. Moeyersons (1976). The differential impact of some soil factors on flow, runoff creep and rainwash, *Earth Surf. Proc.*, **1**, 151–61.

Dissmeyer, G. E. and G. R. Foster (1985). Modifying the universal soil loss equation for forest land. In S. A. El-Swaify, W. C. Moldenhauer and A. Lo (eds), *Soil Erosion and Conservation*, pp. 480–95, Soil Conserv. Soc. Am., Ankeny, IA.

Finney, H. J. (1984). The effect of crop covers on rainfall characteristics and splash detachment, *J. Agric. Engng. Res.*, **29**, 337–43.

Foster, G. R. (1982). Modeling the erosion process. In C. T. Haan, H. P. Johnson and D. L. Brakensiek (eds), *Hydrological Modeling of Small Watersheds*, pp. 297–380, Am. Soc. Agric. Engnrs., Monograph No. 5.

Giroud, J. P. and R. G. Caroll, Jr. (1983). Geotextile products, *Geotechnical Fabrics Report*, **1**, 1.

Herwitz, S. R. (1986). Infiltration-excess caused by stemflow in a cyclone-prone tropical rainforest, *Earth Surf. Proc. Landf.*, **11**, 401–12.

Hudson, N. W. (1981). *Soil Conservation*, Batsford, London.

Laflen, J. M. and T. S. Colvin (1981). Effect of crop residue on soil loss from continuous row cropping, *Trans. Am. Soc. Agric. Engnrs.*, **24**, 605–9.

Lang, R. D. and L. A. H. McCaffrey (1984). Ground cover: its effects on soil loss from grazed runoff plots, Gunnedah, *J. Soil Conserv. Serv. NSW*, **40**, 56–61.

Masounave, J., R. Denis and A. L. Rollin (1980). Prediction of hydraulic properties of synthetic non-woven fabrics used in geotechnical work, *Canadian Geotech. J.*, **17**, 517–25.

Morgan, R. P. C. (1985). Effect of corn and soybean canopy on soil detachment by rainfall, *Trans. Am. Soc. Agric. Engnrs.*, **28**, 1135–40.

Morgan, R. P. C. and H. J. Finney (1987). Drag coefficients of single crop rows and their implications for wind erosion control. In V. Gardiner (ed.) *International Geomorphology 1986 Part II*, pp. 449–58, Wiley, Chichester.

Morgan, R. P. C., H. J. Finney, H. Lavee, E. Merritt and C. A. Noble (1986). Plant cover effects on hillslope runoff and erosion: evidence from two laboratory experiments. In A. D. Abrahams (ed.) *Hillslope Hydrology*, pp. 77–96, Allen & Unwin, London.

Noble, C. A. and R. P. C. Morgan (1983). Rainfall interception and splash detachment with a Brussels sprouts plant: a laboratory simulation, *Earth Surf. Proc. Landf.*, **8**, 569–77.

Quansah, C. (1985). Rate of soil detachment by overland flow, with and without rain, and its relationship with discharge, slope steepness and soil type. In S. A. El-Swaify, W. C. Moldenhauer and A. Lo (eds), *Soil Erosion and Conservation*, pp. 406–23, Soil Conserv. Soc. Am., Ankeny, IA.

Rickson, R. J. and R. P. C. Morgan (1988). Approaches to modelling the effects of vegetation on soil erosion by water. In R. P. C. Morgan and R. J. Rickson (eds), *Erosion Assessment and Modelling*, in press.

Singer, M. J. and J. Blackard (1978). Effect of mulching on sediment in runoff from simulated rainfall, *Soil Sci. Soc. Am. J.*, **42**, 481–6.

Styczen, M. and K. Høgh-Schmidt (1986). A new description of the relation between drop sizes, vegetation and splash erosion. In B. Hasholt (ed.), *Partikulaert bundet stoftransport i vand og jorderosjon*, pp. 255–71, Nordisk Hydrologisk Program NHP Report No 14.

van Asch, W. J. Th (1980). Water erosion on slopes and landsliding in a Mediterranean landscape, *Utrechtse Geografische Studies*, 20.

Wischmeier, W. H. (1975). Estimating the soil loss equation's cover and management factor for undisturbed areas. In *Present and Prospective Technology for Predicting Sediment Yields and Sources*, USDA Agr. Res. Serv. Pub. ARS-S-40, 118–124.

Yair, A. and H. Lavee (1976). Runoff generative process and runoff yield from arid talus mantled slopes, *Earth Surf. Proc.*, **1**, 235–47.

# Urban Land Use

Geomorphology in Environmental Planning
Edited by J. M. Hooke



# 5 Urban Planning Policies for Physical Constraints and Environmental Change

**IAN DOUGLAS**
*School of Geography, University of Manchester*

The tasks of the urban geomorphologist are threefold: to know the ground the city is built upon; to understand the present day geomorphic processes as modified by urbanization; and to predict future geomorphic changes which are likely to result from urban development. Essentially these tasks require knowledge of the past, understanding of the present and the ability to forecast the future.

Urban planning policies generally encompass some knowledge of the past in terms of physical constraints on urban development, but generally pay relatively little attention to the present and the future. In England and Wales, the planning procedure permits consideration of geomorphology during the preparation of structure plans and local plans and in the development control processes when objections to specific planning applications may be raised. An examination of some structure and local plans for NW England, revealed that geomorphology is not mentioned specifically; that landform is used with reference to the shape of the ground surface in a static, rather than a dynamic sense; but that particular geomorphic features are recognized as constraints on development, as environments to be protected or as significant in terms of mineral resources (Table 5.1). Thus, in the Greater Manchester Structure Plan areas of lowland basin peat, upland peat zones, slope and topography, liability of flooding and mining subsidence are all seen as constraints on urban development. Both the Greater Manchester and Merseyside structure plans recognize the significance of fluvio-glacial sands for sand and gravel resources and the special industrial value of the Shirdley Hill sands for the glass industry. The Merseyside and Central and North Lancashire Structure Plans acknowledge the importance of coastal protection, particularly the need to restore areas of damaged or degraded sand dunes along the coasts between the Ribble and Dee estuaries.

The lack of geomorphology in urban policy and planning in NW England occurs against an historical record of many comments on the geomorphic and hydrologic consequences of urbanization. Engels' classic description of Manchester in his *The Conditions of the Working Class in England* (Engels, 1845) contains frequent reference to the siltation and flooding of the Irk

**Table 5.1** Recognition of geomorphology in some plans for NW England

| Area | Type of plan | Date | Geomorphological content |
| --- | --- | --- | --- |
| Greater Manchester | Structure Plan Written Statement | 1977 | Policies relating to derelict land, mineral extraction and proposals for environmental and recreational development within river valley areas |
| | | | Recognition that sand and gravel resource areas should not be sterilized, for although the widespread fluvio-glacial Middle sands and the Aeolian Shirdley Hill sands offer good opportunities for exploitation, there are few areas of open land still available on these formations. |
| Greater Manchester | Structure Plan Report of Survey Openland and Physical Restraints on Development | 1975 | Strategic policy implications of: basin and upland peat soils; slope and topography; land liable to flood; and mining subsidence as constraints on urban development. |
| Mersey Valley (Greater Manchester) | Local Plan Report of Survey and Issues | 1978 | Objective to conserve and enhance the physical landscape of the Valley as an entity. Attention to past and present flood risks. |
| Croal-Irwell Valley (Greater Manchester) | Local Plan Report of Survey and issues | 1979 | Recognition of role of 'Land Form and Geology' in valley character and of value of 'the geology' as 'an educational and interpretative resource'. Need to identify areas of natural history and geological interest which should be safeguarded. Recognition that the natural forces which formed the river 'have now been modified or negated by man, in his construction of drainage systems, sewers, weirs and reservoirs. The river is confined in places by reinforced banks, erosion is reduced, tributary valleys filled, river terraces raised by tipping and new valleys created for railways and motorways. Value of gorge sections as scenic resources. |

| Area | Plan | Year | Content |
|---|---|---|---|
| Medlock Valley (Greater Manchester) | Subject Local Plan Draft Written Statement | 1981 | Proposal to conserve and develop a variety of habitats and landscape character throughout the valley in order to provide a choice of scenery for different types of recreation activity and a richer experience for visitors to the valley. |
| | | | Proposal to introduce recreation to newly restored areas only after the landscaping has become established in order to prevent erosion of 'an immature landscape'. Specific proposals with high landform change content include reclamation of derelict land including partial filling of former quarry to provide open space incorporating view point at Glodwick Lows. |
| Merseyside | Structure Plan Written Statement | 1979 | Policy that planning authorities will normally refuse permission for the winning of aggregates from beneath quality farmland, sand dunes and foreshores. However, policy also allows extraction of sand for special industrial uses on the foreshore except where extraction is shown to be actually or potentially deleterious to coastal defences, foreshore quality, nature conservation or the shellfish industry. 645 ha of Shirdley Hill sands used for glassmaking may still be extracted, only a thin layer being required and restoration of the high quality farmland being possible. |
| | | | Policy to restore areas of damaged or degraded sand dune at Ainsdale, Formby Point, Hightown, West Kirby Hoylake and Wallasey. |
| Merseyside | Structure Plan Report of Survey Natural Resources | 1979 | Particular attention to erosion of the sand dunes of the Sefton Coast. Specific discussion of impact of removal of moulding sand from coastal deposits near Southport and on need to regulate development over Shirdley Hill sand deposits. |
| Lancashire | Central and North Lancashire Structure Plan Explanatory Memorandum | 1981 | Policy to support measures of coastal protection and flood prevention which safeguard communities, farmland and public services. |

and Medlock. Many major floods on these tributaries and on the Irwell and Mersey occurred in the nineteenth century (Bracegirdle, 1971; Massey, 1976). So great was the siltation of the Irwell, that half the channel capacity of the river between Salford and Manchester was lost, flood heights increased dramatically and, in 1868, the government appointed a Commission to enquire 'into the best means of preventing pollution of rivers (Mersey and Ribble Basins)'. The Royal Commission's report contains some excellent early examples of the practical value of analyses of the hydraulic geometry of rivers (Douglas, 1985a).

Despite this background, the local plans for Manchester river valleys show far less awareness of geomorphology than would be expected in planning studies of fluvial landscapes. The Croal–Irwell Local Plan Report of Survey recognizes the role of 'land form and geology' in the character of the valleys and the value of 'the geology' as 'an educational and interpretative resource'. The plan acknowledges a need to identify areas of natural history and geological interest which should be safeguarded. However, natural processes are not stressed because their work has been 'modified or negated by man in his construction of drainage systems, sewers, weirs and reservoirs'. Lack of perception of how the river has changed as a result of urbanization and how the natural processes respond to these changes may explain why the Croal-Irwell document fails to mention past flood hazards in the Lower Kersal area and the slope instability and subsidence problems of The Cliff (Douglas, 1985a; Harrison and Petch, 1985). Despite the record of 1946 flooding and the existence of a North West Water Authority flood warning scheme (Noonan, 1986), a new private development was allowed to proceed in 1977 in the area liable to inundation only to be flooded in 1980.

The Mersey Valley Local Plan Report of Survey and Issues notes the dissection of the floodplain by big meanders near Urmston, but fails to explain the dynamic nature of these meanders immediately downstream of the end of the channelized reaches of the river. In a discussion of the problems of rapid channel change and meander cutoff development of this section of the Mersey, Pearson (1947) asks 'Why is it that, in connection with the present spate of Town Planning talk and literature, rivers, the main fixed and fundamental feature of any planning, and often the main feature of green belts, are rarely mentioned?'

This rapidly changing reach of the River Mersey illustrates how several interrelated urban policy issues are intimately involved with geomorphology. Increased recreational use of the floodplain, including the abandoned meanders and newly formed sand bars is envisaged by the Mersey Valley Local Plan. Much of the land already owned by Trafford Metropolitan Borough Council has been developed as playing fields, the edges of which are threatened by erosion. The Borough Council is concerned at the loss of land, bank retreat of over 2.5 m occurring during a single storm on 30 December 1986, and a similar quantity at another location during a subsequent storm on 25 June 1987. However, safety considerations are also involved. If the policy is to encourage greater public use of floodplain land for sport and recreation, can visitors be allowed to wander close to banks subject to active slumping where slip surface cracks are visible? Answering this policy question clearly demands consideration of future trends in bank erosion, of the stability of any modified bank and of the possible effect of vegetation growth.

Other policies on environmental and ecological issues suggest that a downstream extension of the channelization would be undesirable, that abandoned meanders have ecological values as wet land habitats and that wildlife on the floodplain should be encouraged. The Local Plan also suggests that unnecessary waste disposal by landfill tipping in the valley should be eliminated, yet tipping operations can be highly profitable commercially. Sometimes filling operations can be seen as a means of increasing channel stability. Prior to local government reorganization in Greater Manchester 1974, many authorities filled parts of the floodplain and thereby reinforced the artificial levée banks. For example, household refuse dumped on Stretford Ees, 4 km upstream of Urmston, between 1923 and 1949 raised the surface level 1 m above the embankment, increasing flood protection for the main A56 road to Altrincham (Massey, 1976). Such fill is still occurring on the floodplain, but not close to the river, at the Greater Manchester Waste Disposal Authority's Stretford Civic Amenity Site, north of the M63 motorway, and in an area to the south of the motorway further upstream. However, at the Ackers farm private tip, Carrington Lane, south of the river near Flixton the dumped material actually forms part of the river bank. Here planning applications to extend the tip, with further building-up of the river bank require the Metropolitan Borough to clarify its policy on the future of the meandering river. Any local reinforcement of the bank may induce additional erosion at another site further downstream and would force floodwaters to spread over less well protected parts of the valley floor.

The Mersey River in Greater Manchester illustrates the ramifications of geomorphic processes into many fields of urban policy, into recreation provision, environmental planning, transportation, waste disposal and flood protection. Nevertheless, study of planning documents for this part of England shows that the place of geomorphology in urban policy is patchy and incomplete. Even though the former Greater Manchester County developed a coherent policy for river valley improvement, such policy was far more influenced by the biological sciences that the earth sciences. The plans give far more attention to flora and fauna, industrial archaeology and historical sites, rather than to either sites of particular geomorphological interest or the risks and future problems that may arise through the work of geomorphic processes. Geomorphology is seen as a background factor rather than as a dynamic element in the landscape.

To assess how geomorphology comes to play a background role in urban policy, the manner in which policy comes to take account of geomorphic problems needs to be assessed. Generally, urban policy-making and planning bodies have few, if any, provisions for the formal involvement of professional geomorphologists. Planning departments often have an earth scientist as minerals officer, but such a person is usually a traditionally trained geologist with little geomorphology. Given this lack of scientific advice about geomorphology at the formal stage, the impact of geomorphic processes usually comes when they cause, or threaten, a disaster.

## THE NATURE OF URBAN POLICY RELATED TO GEOMORPHIC PROBLEMS

Much of the impact of earth processes and geomorphic phenomena on urban policy comes only after disasters, sadly often when there has been tragic loss of life. In this sense, policy making is a response, not primarily to scientific evidence of possible future problems, but to disaster. This type of policy response, common in many countries, is well illustrated by events in Australia. Extremes of flood and drought are such a feature of Australian life that a balladeer once wrote:

> Queensland, thou art a land of pests!
> . . .
> And then it never rains in season
> There's drought one year and floods next season.

Government policy to locate urban development outside flood-prone areas was pronounced as early as 1819 by Governor Macquarie who criticized new settlers for their 'willful and wayward habits of placing their residences within the reach of floods, and in putting at defiance that impetuous element which it is not for man to contend with'. Such foresight has not always been matched by twentieth-century state and federal governments in Australia. The typical response by governments to a flood event in Australia (Douglas, 1979) is, in chronological sequence:

(1) Emergency assistance.
(2) Financial provision for flood relief grants and loans.
(3) Pressure by local governments on state and federal authorities for flood mitigation works.
(4) Establishment of government departmental or inter-departmental committees.
(5) Some consultation with local government.
(6) Preparation of a report describing flood mitigation works or evaluating multipurpose water management schemes.

(7) Possible policy changes through an Act authorizing work to proceed or introducing new land use regulations, *or* no further action.

Following the severe flooding of the early 1970s, the New South Wales Government broke out of the traditional pattern of localized flood mitigation work in response to specific flood events to introduce a statewide policy restricting investment of government funds for building on flood-prone land. The 1977 Directive required that no building funded by the government should be constructed within the 1:100 year flood hazard zone. The extent of such zones was unknown for most Australian towns, and in many small country towns

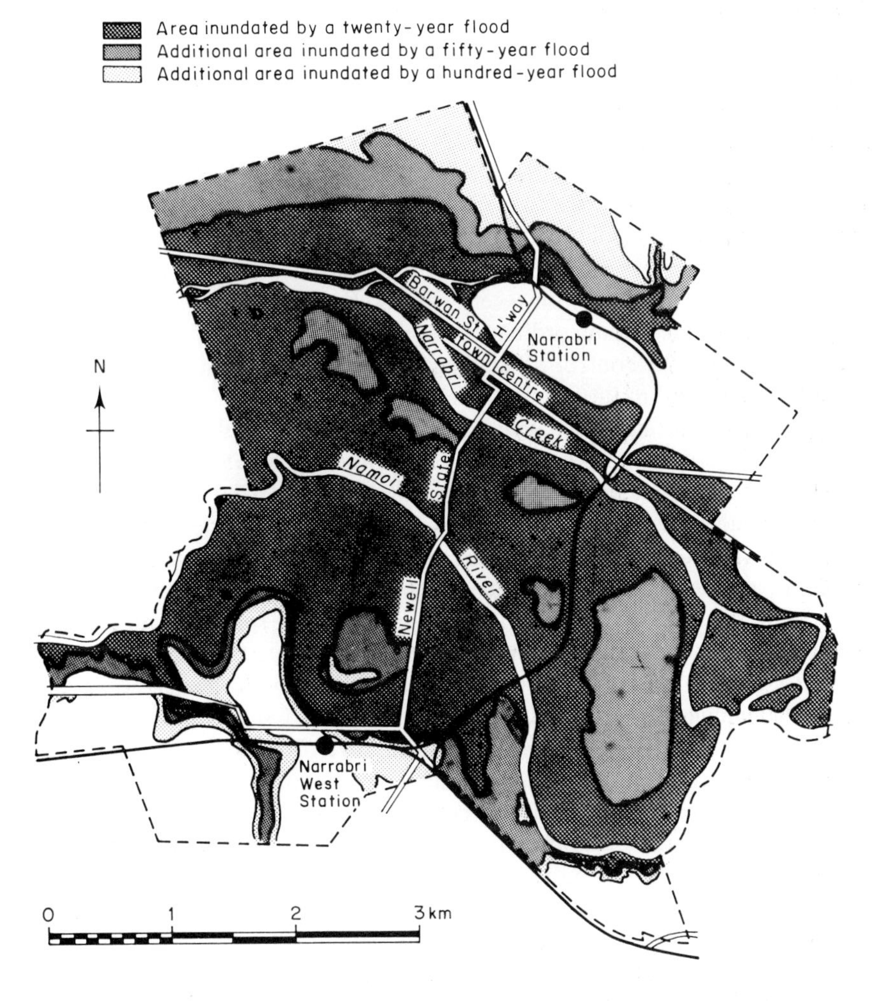


**Figure 5.1** Floodplain map to show probability of flood hazard at Narrabri, New South Wales (based on work of NSW Water Resources Commission)

such as Narrabri (Figure 5.1) covered virtually the whole urban area. Inevitably, there were great problems for urban property owners, developers and planners as a result of this policy. Initially, the policy was to produce floodplain maps of individual towns. After a few years the policy changed, no more maps being issued. Data on flood-prone areas are still filed in appropriate ministries, but the publication of maps had been found to cause hardship to some property owners and was no longer politically acceptable. Policy containing geomorphological evidence was thus only acceptable while flood risks were higher on the political agenda. In the economically lean drought years of the early 1980s, when few floods occurred, property values had greater political significance than flood potential. Thus, perhaps, just as extreme events are of great significance in landform evolution, so they are in the development of urban policy relating to geomorphic processes.

## MASS MOVEMENT EVENTS AND POLICY DEVELOPMENT

The state and county policies for control of building on flood-prone land and landslide hazard areas in California are a response to past disasters and have been strengthened after more recent events. For example in the City of Los Angeles, the initial grading ordinance developed after severe storms in 1951 and 1952 focused on the control of grading through permits and inspections, and on specific design limitations of slope steepness, drainage provisions, and fill compaction. These controls have been strengthened in subsequent years. For example, cut slopes were limited to 1:1 gradients in 1952; after the 1957–8 storms, the maximum angle was reduced to 1.5:1 in the Santa Monica formation and in certain other areas; and in 1963, the maximum angle of most cut and fill slopes was reduced to 2:1 (Cooke, 1984). This pattern of evolution of policy for development control on unstable slopes is paralleled in many other countries.

The rapid growth of Hong Kong on its confined site spreading up colluvium-mantled, deeply weathered granitic slopes has been tragically affected by several serious landslide events, especially those associated with the severe storms of June 1960 which left 64 dead and 2500 homeless, in Fat Kwong Street in 1970, and at Sau Mau Ping and Po Shan Road in 1972 (So, 1986). The government response, aware of the lack of detailed geomorphological and geological mapping, was to establish the Geotechnical Control Office which since 1977 has produced terrain classification, geotechnical land use and engineering geological maps which are used to control development and to provide guidance on individual construction projects. Geomorphological information is now a part of the development control process in Hong Kong.

Much geomorphic information is now collated and incorporated into planning schemes in various municipalities in the USA. The Santa Barbara County Planning Department, sensitive to the earthquake and landslide risks of southern

California, uses a geologic problem index (GPI) to rate geologic and geomorphic hazards for a given area on both an individual and collective basis. The system rates individual hazards such as seismicity, location of active faults, tsunami flood potential, liability to liquefaction and landslide risk for each 2 ha grid square as high, moderate or 'none to low'. GPI values are obtained by multiplying each geologic and geomorphic hazard factor by a weighting factor which expresses the seriousness of the hazard, its frequency of occurrence and the difficulty of alleviating it. Using GPI index maps the County Planning Department makes recommendations concerning land-use planning, subdivision procedures, grading codes, building codes and land-stability insurance. For example, it is recommended that areas classified as severe on the GPI index:

> Should be given primary consideration for minimum development and use. They could be planned as natural areas, or for recreational or agricultural use. If development is permitted, it should generally be of low density.

The whole Santa Barbara scheme is seen as a good example of how geomorphic and geologic information can be collated, evaluated, digitized, rated and used in the development of intelligent plans for land use (Ziony, 1985). However, such earth science inputs to urban policy must be seen in the wider context of decision making. As Ziony and coworkers (1985) point out, planners, engineers and decision makers live and work in a complex environment. Even in a seismically active area like the Los Angeles region, the geomorphic and geologic environment is just one aspect of their life and work. Social, economic, aesthetic and political considerations may be more important or more readily apparent to them. Any use of earth science information and the reduction of geologic and geomorphic hazards must depend on such factors as:

(a) Continued awareness and interest by the public and their decision makers.
(b) Meticulous updating of hazard information and maps by geologists, seismologists and geotechnical engineers.
(c) Careful revision of enabling legislation (if needed) by legislative bodies.
(d) Accurate site investigations by registered geologists or geotechnical engineers.
(e) Conscientious administration of regulations by inspectors.
(f) Consistent enforcement by government officials.
(g) Sustained support of inspection and enforcement officials by political leaders.
(h) Judicious adjustment of regulations by administrative—appeal bodies.
(i) Skilful advocacy by public officials and informed interpretation by the courts, if the regulations are challenged.
(j) Concern for individual, family and community health, safety and welfare by home-buyers and real-estate developers.

Policy formulation and amplification as a response to disaster is not always adequate. The need for dwellings, for shelter for growing urban populations often outstrips both the ability of urban policy-makers to incorporate geomorphological information, and the ability of the authorities to ensure enforcement of such policies as do exist.

## EVALUATION OF HOUSING DEVELOPMENT SITES

Housing, or shelter, is a universal need, yet homelessness and inadequate shelter remain severe problems throughout the world. For those fortunate enough to be able to invest their savings in a house or pay a large part of their income to rent accommodation, damage to or loss of their shelter by geomorphic processes is a severe extra penalty which ought to be avoided. Those less fortunate find the shortage of housing and land in many rapidly growing cities so severe that they have no option but to live in spontaneous squatter settlements, where they have to accept the risks of living in unsafe, hazardous sites, such as those affected by the well-known landslide problems of Rio de Janeiro (Mousinho de Meis and da Silva, 1968; Jones, 1970). However, examples of planned, officially approved, government and privately financed housing schemes which are damaged by geomorphic processes abound. They occur not only in environments which are tectonically active and prone to geomorphic instability or climatically difficult, but also in areas regarded as stable and well-mapped topographically and geologically. While some gaps in information, or the use of information in planning and development control are probably always going to exist, the adequacy of urban planning policy must be questioned when available information and expertise is not used.

Two private housing developments in the valley of the River Irwell in NW England illustrate this point. At Lower Kersall on the Irwell floodplain, severe flooding occurred in 1946, following approximately the same pattern as a major flood in 1866. A large council house estate was inundated, and several measures, including channel enlargement and an artificial meander cutoff were undertaken in subsequent years. However, in 1977, planning permission was given for a small private housing development in St Aidan's Close, adjacent to the previously flooded council estate. In October 1980, the Irwell experienced another flood as large as that of 1946 and the new houses were flooded. Enquiries by Salford City Council revealed a lack of full information flow between officers of the Council and those of the North West Water Authority. None of the house purchasers appeared to have been warned of the flood risk when they bought their properties.

The Irwell rises in the Pennines and descends towards Lower Kersall and Manchester from Rossendale through a valley enlarged by glacial meltwaters spilling over from the icesheet in the Ribble Valley to the north. The slopes of this part of the valley, like those of many other valleys in this part of England, are mantled with glacial till. Near Haslingden, construction began on a housing site on the valley side at Ewood Bridge in 1973. Some 66 houses had been built, sold and occupied or were partially completed when movement of some floor slabs occurred and cracks appeared on some walls. Construction work was stopped and the resale value of the houses fell dramatically. Aerial photographs taken in 1976 show two landslip scars running across the site. The landslips probably occurred 12 000 to 10 000 years ago during post-Glacial climatic amelioration by movements along the interface between an underlying laminated clay till and an overlying boulder clay. Whether the housing construction and associated excavation and filling operations reactivated these slips or merely accentuated local instability is unclear. However, after reports from several geotechnical consultants, Rossendale Borough Council concluded in 1980 that while the existing structures could remain, there was too high a risk of instability for any further construction to be permitted.

In both these Irwell Valley cases, geomorphological information was not used at the appropriate time. Valley side landslips were well known, from the experience of engineers building railways and reservoirs in nearby Pennine Valleys in the nineteenth century, and from the more recent construction of the M62 motorway. The flood history of the Irwell was thoroughly documented by the water authorities and in every local history library. In failing to bring geomorphic problems into the development control procedures, urban policy demonstrates a lack of awareness or consciousness of earth surface processes and the nature of local landforms.

## POLICY FOR URBAN LAND CAPABILITY ASSESSMENT

Similar post-construction geomorphic problems saw the gradual evolution of a new policy towards evaluation of housing sites in New South Wales. The construction of new housing estates on deeply weathered granites on the low rounded hills near the city of Bathurst led to severe soil erosion. After being called in for remedial work, the Soil Conservation Service developed techniques of pre-development land appraisal, that became adopted as policy by many municipal authorities seeking to avoid severe erosion, flooding and expansive soil problems.

The Soil Conservation Service of New South Wales (Hannam, Davis and Cocks, 1986) has suggested that a total catchment management strategy may be incorporated into planning policies through the environmental plans provision of the State planning system. The 1979 Environmental Planning and Assessment Act allows proposed changes in land use to be assessed and either approved (with or without conditions) or prohibited from taking place. The Act provides for environmental control through Local Environmental Plans, Development

Control Plans and Regional Environmental Plans. When a catchment being planned falls wholly within one local government area, a single Local Environmental Plan could be used to control land use. Regulation and enforcement at the local level can be achieved through the Development Control Plans.

The first step in implementing the concepts of total catchment management in urban areas is to compile an inventory of the environmental resources, of each catchment, in which urban development is to proceed. The inventory can then be used as a basis for assessing the capability of each parcel of land for urban uses. From the capability assessment, future development recommendations able to maintain the long-term stability, productivity and other values of that catchment's environmental resources can be formulated. These parameters are provided in the New South Wales Soil Conservation Service's Urban Land Capability Classification System (Hannam and Hicks, 1980). Firstly, it provides an inventory of the physical environmental characteristics which are relevant to urban uses. It then determines the land's capability for urban uses and outlines appropriate associated land mangement measures for each urban capability class. These measures include the implementation of soil erosion and sediment controls as well as stormwater management systems. Such measures assist in maintaining the long-term productivity and stability of an urban catchment. The next step towards implementing Total Catchment Management involves the integration of all urban land use demands with urban capability information, to provide recommendations for a final set of land uses within an urban catchment (Johnston, 1986).

The NSW Soil Conservation Service carried out a variety of urban land capability studies in the 1970s, completing a computer-based inventory, based on a 16.1 ha grid cell for the Bathurst–Orange growth area in 1978. Landform, soil, slope, soil erosion, land use and climatic information is used to derive urban land use capability assessments. The report to the Bathurst City Council (Hannam, Emery and Murphy, 1978) advocated a catchment management policy to regulate development in several catchments, whose streams flowed through the urban and commercial aras of the city. The soil conservationists recognized that large scale urban development in the catchments could cause higher peak discharges and sediment loads, producing channel changes which would affect the older urban commercial areas downstream.

## POLICIES TO REGULATE THE OFF-SITE IMPACTS OF URBAN CONSTRUCTION

Housing site development changes runoff water quantity and quality downstream, often producing channel changes, siltation, scour of bridge abutments and widespread, frequent nuisance flooding. Such problems are well

**Table 5.2** US soil conservation guidelines for urban areas (as used on information sheets distributed by Colorado SCS offices)

A *Conservation tips for builders*

1. Choose a site that has good natural drainage, topography and soils.
2. Locate the building on the site so there is adequate drainage away from it.
3. Plan driveways to avoid excessive grades.
4. Hold site grading to a minimum.
5. Save trees and other existing vegetation.
6. Expose as small an area of land for as short a period as possible.
7. Plant temporary vegetation during development in critical areas.
8. Install conservation measures to protect the site.
9. Plant permanent vegetation as soon as practical after construction.
10. Repair and maintain conservation practices.

B *On construction and conservation in urban areas.*
*Conservation starts when construction starts: a ground plan outdoors is as important as a floor plan inside.*

(A) *During construction*

1. Disturb only the area necessary for construction.
2. Protect trees and shrubs that are to be retained.
3. Stockpile separately and protect valuable top soil for later use as lawns and in flower beds.
4. Protect the site while construction is underway. Bare soil encourages runoff and erosion. Complete the final grading, seeding, and sodding just as soon as possible.
   (a) Protect areas such as stockpiles of soil and rough graded areas by applying a mulch. A wood fibre slurry used as a mulch applied mechanically is an excellent way of protecting an area until final grading is done. Straw chopped and blown into place can be used. A light application of asphalt spray will hold it in place. On small but critical erosion areas, a straw mulch can be held in place with a coarse fibre netting.
   (b) When stockpiles and rough graded areas must remain for an extended period of time before final grading can be done, establish plant cover of small grain, grass, or legumes, or combinations of all three. Here an excellent seeding method is the use of the combination mulcher and seeder which sprays mulch, seed, and fertilizer onto the area to be seeded as a slurry in one operation. Seeding and spreading of fertilizer can be done by hand without seedbed preparation when the area has a cloddy, loose, and moist surface. This is often necessary on small areas inaccessible to mechanical equipment. Ordinary farm equipment such as grain drill, cultipaker, and tillage tools are useful where machinery can be operated over the area to be seeded.
5. Control runoff water
   (a) Divert away from critical erosion areas.
   (b) Create temporary water impoundments and silt traps, using earth dams and/or dugouts, when the topography is suitable. Valuable soil will be saved and runoff water will be unable to damage adjoining property.

*(continued)*

**Table 5.2** *(continued)*

(B) *Complete final grading, seeding and sodding as quickly as possible*

1. Remove all debris such as tree stumps, scrap lumber, mortar or concrete, and rocks. Do not bury them; wood will eventually rot and cause settling; rocks, mortar and concrete can cause real difficulties in lawn maintenance and later construction.
2. Grade to provide drainage of the developed area. This may require practices such as water courses and diversions. Special water controls may require detailed design and layout.
3. Apply necessary fertilizer.
4. Prepare adequate seedbed.
5. Establish permanent vegetative cover by seeding or sodding. Use recommended grasses or legumes.
6. Erodible areas when seeded should be mulched to prevent excessive erosion and loss of seed.

recognized in many countries, and are well addressed by the manuals on urban erosion control, prepared by the US Soil Conservation Service (USSCS) for many of the United States. Urban policies often adopt guidelines similar to those proposed by the USSCS (Table 5.2).

In tropical cities where frequent, intense rains and deeply weathered surface materials combine to produce high rates of urban erosion during construction (Gupta, 1987; Douglas, 1985b), the runoff and sediment yields from building sites are subject to control by urban bye-laws. In Singapore, the Surface Water Drainage Regulations, 1976, promulgated under the Water Pollution Control and Drainage Act, 1975 (Act 29 of 1975) provides for a Code of Practice on surface water drainage, requires the submission of plans for any such works, and gives the Director of the Department of the Environment the power to require any person, causing the capacity of a drain to be inadequate, to carry out appropriate improvement works within a specified time. The Code of Practice (Singapore Drainage Department, 1978) requires the establishment of a system of drains capable of carrying the runoff from a one in five-year storm over any construction site (Gupta, 1987).

The Singapore Drainage Department recognizes that enforcement action may have to be taken to ensure that contractors do not obstruct flow in existing water courses, disrupt or disturb existing overland flow, or discharge silt into existing watercourses (Yap, 1985). In 1984, about 650 notices were served on contractors leading to 300 convictions and fines of $S 230 000. To try to avoid downstream flooding and siltation, building contractors and consulting engineers and architects are issued with guidelines on silt control and drainage (Table 5.3).

In Kuala Lumpur, the downstream consequences of urban development are well-documented (Leigh, 1982; Douglas, 1978, 1985b). The Kuala Lumpur Draft Structure Plan recognizes that the city experiences, both river flooding and storm water flooding, the latter being due to:

**Table 5.3** Guidelines on silt control and drainage in Singapore (Yap, 1985)

(a) Provide adequate silt-control facilities including lined temporary perimeter drains with minimum 1 metre width of close turfing and adjoining both sides of that cut-off drains, silt-trapping devices, turfing, wash-bays incorporating proper discharge outlets, etc., to control silt and mud from site.
(b) Maintain regularly all such silt-control facilities to ensure their effectiveness and desilt all affected watercourses until completion of the development works.
(c) Ensure that runoff within, upstream of and adjacent to the site can be effectively drained away without causing flooding within the site or in areas outside the site.
(d) Ensure that discharge consequent to de-watering from basement or trench excavation is channelled into silt trapping devices before reaching existing drains.
(e) Implement adequate measures, including the provision of proper barricades between the work areas and existing drains, to ensure that construction materials are not discharged or washed into the drains.

(1) Upstream development.
(2) Downstream construction restricting flow.
(3) Inadequate floodplain storage.
(4) The permitting of development in flood areas (Dewan Bandaraya, 1981).

The plan notes that the streams in the area had in 1978 a total of 281 culverts, 87 per cent of which are inadequate for a one in two-year flood and 285 bridges, 67 per cent of which are inadequate for a 1 in 2 year flood. Policy has therefore concentrated on improving the efficiency of the drainage system.

Although the urban drainage design, standards and procedures for peninsular Malaysia stress that 'natural drainageways should be used for storm runoff waterways wherever possible' and that 'channelizing usually speeds up the flow, causing greater downstream peaks' (Lewis, Cassell and Fricke, 1975), channelization is the adopted policy in Kuala Lumpur. Every new construction site, now increasingly close to the headwaters of small tributaries, is drained by concrete-lined channels which feed urban runoff rapidly into major streams, often through bridges far smaller in cross-section area than the upstream channels, and frequently into unmodified natural channels which erode rapidly. The policy is causing severe hardship for many riparian residents and landholders. It is difficult to relate costs of damage in affected areas down valley to specific land use and drainage modifications in headwater areas.

Although not as dramatic as tropical events in either magnitude or rate of rise, floods and channel changes downstream of urban areas require attention throughout temperate countries. In central England, near Sheffield, the River Rother drains the Chesterfield urban area and neighbouring mining towns. Downstream of Chesterfield, the village of Killamarsh suffered flooding from both overflow of inadequate sewers and backing-up of the River Rother. Cessation of opencast mining along the valley downstream provided an opportunity for Sheffield City Council to create a country park as part of its urban recreation policy. Geomorphological criteria were involved in the creation of a recreational lake, channelization of the river, and stabilization of colliery waste. Flood, dereliction, drainage, waste disposal and recreational demand problems, all essentially the off-site consequences of urban growth and raw material demands, were successfully alleviated by the co-operation of public authorities, nationalized industry and consultants to provide benefits for many different communities (Carter *et al.*, 1984).

These downstream effects of city growth are analogous to many other off-site consequences of urban policy. In the Venice lagoon, vigorous industrialization policy on the mainland led to excessive groundwater abstraction which contributed to the subsidence affecting the island cities. Some off-site consequences ought to be relatively easy to predict, at least qualitatively, such as changes in sediment movement and beach nourishment following harbour works, or the downstream consequences of urban policy associated with buried geomorphic features, such as former late-Quaternary river channels, lake beds, buried karst and permafrost.

## PALAEOHYDROGEOMORPHOLOGY AND URBAN POLICY

Groundwater is often intimately related to geomorphic history, gravel lenses in old river channel deposits providing favourable aquifers, while clay layers between Pleistocene sands create aquifers whose piezometric head may be higher than local phreatic water tables. With groundwater levels sensitive to pumping rates, drainage and changes in infiltration rates due to alterations to surface ground cover, groundwater management requires sound knowledge of subsurface materials and their geomorphic history.

Two deltaic cities, Amsterdam and Bangkok, illustrate different aspects of the problems related to palaeohydrology and Quaternary geomorphology. The ground beneath Amsterdam comprises a series of alternating water-bearing sand strata and impermeable clay and peat layers (Kremer, 1985). Three layers of sands deposited in cold climates are separated by Allerød age clayey sand and Eemian clay, creating confined aquifers. Drainage of large low-lying polders close to the city had reduced the piezometric level of these aquifers, but following a decrease in the pumping rate, water levels began to rise. In one new housing area on a polder on the edge of the city, the underfloor spaces of new homes soon became damp because water was seeping out of the upper sand layer, whose piezometric level was 2 m above the water table in the polder. All regulations and policies had been followed in house construction, but those policies did not take into account the influence of Quaternary sediments on groundwater.

Bangkok has grown so rapidly across the Chao Phya deltaic plain, that inadequate municipal water supplies have had to be supplemented by widespread private groundwater abstraction. The pattern and rate of subsidence caused by pumping is partly determined by re-activated faults within the deltaic material and by the distribution of the highly compressible Bangkok clay, which grades into former beach deposits (Rau and Nutalaya, 1982). The geomorphology of former shorelines, tidal flats and deltaic distributary channels is thus a major factor in the pattern of subsidence and ought to be considered in developing policies to alleviate ground surface lowering.

## KARST AND URBAN POLICY

Four broad categories of problem affect urban areas in karst terrain, those related to the structural stability of the bedrock, those associated with the stability or subsidence of fill in or over karst terrain, those related to foundation engineering on buried pinnacle karst and those related to the contamination of groundwater. Structural failure in karst is due to the collapse of surface or near surface bedrock and is usually expressed in the development of sinkholes. The urban-related processes leading to sinkhole development include:

(1) Lowering of karst water levels by groundwater pumping, leading to withdrawal of support in solution cavities;
(2) Concentration of urban runoff into zones of infiltration at the edges of paved surfaces or structures;
(3) Increased ground loading from buildings or landfill;
(4) Changes in drainage patterns resulting from stormwater drainage, erosion control and sewerage works.

Fill in or over carbonate rock terrain is often liable to rapid settling, as has occurred at several housing sites on infilled former quarries and pits in the chalk around London. Fill over cavernous limestone becomes unstable if fine particles are washed out of the fill into the karst systems below. Former tin mine ponds in the alluvium covering the buried karst beneath part of Kuala Lumpur have been dewatered, before being filled and reclaimed. However, the fill is often washed into the karst system, leading to damage, and even collapse, of the newly built properties above (Tan, 1986).

Construction on buried karst becomes a problem when tall buildings require deep foundations, which penetrate through the overlying stiff clays and sands of the alluvial plains. In humid tropical areas, major problems are caused by the pinnacled surface of the karst. Differences of 10 to 20 metres between the summits of pinnacles and the base of adjacent depressions may mean high foundation costs. Furthermore, the underlying limestone may only be a few metres thick, with a large cavern below. Traditional site investigation by drilling provides only limited data, for even with 150 to 200 drill holes per site, the possibility that some of the piles used to support a high-rise building would be over a large karst cavity, cannot be completely eliminated (Tan, 1986).

Contamination of groundwater by untreated waste water, leachate from landfill, stormwater runoff and accidental spillage is a problem in all karst terrains. Wastes washed into karst hydrologic systems often travel to springs and water sources quickly. Several Yugoslav urban typhoid fever epidemics were caused by contamination of spring waters with faecal substances after heavy rains (Pokrajcic, 1976). Leachates from poorly insulated sites can pass through karst systems into nearby rivers. Often the health risks are not as great as the visual deterioration of discoloured streams suggests (Turk, 1976), but much depends on whether occasional storms wash accumulations of noxious substances into channels in high concentrations.

Policies to cope with these problems are often inadequate, particularly as the nature of karst hydrogeology and geomorphology is not well understood by decision makers. China has several urban planning policies to limit the impact of urban development in karst areas. Regional planning in karst areas should:

(1) Categorize areas according to severity of solution phenomena;
(2) Develop appropriate building and waste disposal standards;
(3) Plan land uses that can tolerate some structural failure and which avoid groundwater contamination (Sheedy *et al.*, 1982).

Ideally, detailed surveys of karst terrains should form a basis for planning urban development, but all too often they are lacking.

In Britain, karst phenomena are particularly important in the rapidly expanding housing developments on the Jurassic limestones and Cretaceous chalk. To the south of London, suburbs spread up the chalk dip-slope of the North Downs. In places old chalk pits and quarries that were filled with rubbish decades ago are used for building land, often without the developers realizing the change in the substrate. As water moves down through the fill, especially where it is concentrated by drainage from paths and roadways, it leads to solution of some material and settling and compaction of the remainder. Damage to houses as a result of such subsidence is often reported, but land use policies take inadequate account of the subsurface materials and irregularities of chalk terrain.

The variety of types of karst and of geomorphic situations occupied by urban areas is so great, that levels of understanding and awareness of karst problems, both among engineering and planning professionals and among lay legislators, and the general public differ markedly from place to place. Consequently, statutes and legislation concerning the impact of karst processes, especially sinkhole development, vary in effectiveness and interpretation (Quinlan, 1986). Considerable attention has to be paid to the criteria for establishing the precise nature of the geomorphic processes which have caused, or may be likely to cause damage in an urban area.

The need for precise criteria is well illustrated by the insurance rules in Florida, which provide for a pool concept or participation plan in which insurers are required to provide cover for insurable sinkhole losses if requested by landowners. Geomorphologists have become involved in the legal interpretation of these rules. Salomone (1986a) suggests that the statement about sinkhole risk reading 'only when such settlement or collapse results from subterranean voids created by the action of water on limestone or similar rock formations' should be amended and augmented to end 'on limestone or other soluble rocks, or from the movement of overlying sediments into such voids'. In the lowlands of Florida, careful assessment of geomorphic processes is essential in alleged cases of settlement caused by sinkhole collapse. In one typical case, settlement was due to compression of peat by the structure itself, rather than to subsurface solution processes (Salomone, 1986b). In a country where policy implementation often evolved through a series of legal precedents, such involvement of geomorphologists in court cases helps to provide urban policy with a better earth science basis.

## CONCLUSIONS: TOWARDS A DYNAMIC URBAN POLICY

Urban policy related to geomorphology has been largely responsive, being developed and amended after events have shown a need for policy to avoid future repetitions of loss or damage. In this way, urban policy is preventative, or limiting, rather than creative and prospective. This may stem from the general perception, prevalent in urban planning literature, of geomorphic features and process as part of a general set of physical limitations on urbanization. Geomorphological criteria appear as control factors in building regulations, construction guidelines and environmental policies.

Many urban authorities have moved from simple limiting policies to incorporating geomorphological information in the planning process. Some even act as geomorphological agents, reclaiming derelict land, re-designing drainage networks and creating new landforms. Specific examples include the landscaping at Warrington New Town (Tregay and Moffatt, 1980); the use of stormwater retention ponds and flood basin as recreational lakes in Manchester (Douglas, 1985a) and Winnepeg (Hough, 1984); the planning of new suburban communities, such as Woodlands 40 km north of Houston, Texas with such objectives as design for maximum recharge, protection of permeable soils, retardation of erosion and siltation and increase of stream baseflow (Hough, 1984); and the incorporation of earth science information in the planning process as in Santa Barbara, California (Ziony, 1985) and Suez City (Cooke *et al.*, 1982). More generally, scientific government agencies are vigorously arguing the case for more use of earth science information in urban planning. In Britain, the Department of Environment has funded many thematic mapping programmes in urban areas by the British Geological Survey. The report on geological surveying in Britain (Butler, 1987) emphasizes the need for earth science data 'in many civil engineering projects' stressing that 'reliable geological information at hand during the critical early stages of a project, whether it be a large hydroelectric system, a nuclear power station or a housing development, can significantly shorten the development time and also reduce the risk of subsequent, untoward and very costly events'. However, the thematic mapping of urban areas is seen as part of the responsive, as opposed to the core programme, continuation of which is encouraged. With the growth of a national geologic database and further thematic mapping, more earth science information should be available to help form urban policy. Whether it will include all the geomorphology that British earth scientists could provide is doubtful due to the relatively narrow view of earth sciences held in British geological circles.

By contrast, the more open outlook of geoscientists in the United States has led to vigorous advocacy of the use of earth science data in urban planning, with many schemes for evaluating geomorphic and hydrologic hazards in and around cities (Robinson and Speiker, 1978; Blair and Spangle, 1979; Helley *et al.*, 1979). Several authorities, such as those in Santa Barbara mentioned earlier and Oregon and Massachusetts have planning and development control procedures using soil conservation information on maps and in databases. The inputs from state and federal geological and soil surveys should be complementary, but it is not clear how much collaboration there is between the two services. In West Germany, the problem ought to be overcome, as soil survey there is part of the geological survey, which ought to mean a single channel of earth science advice to urban policy makers.

Whether or not the geomorphological advice and information is available in a suitable form, its use will depend on the perceived priorities of local communities, planning officials and government representatives. Action in the light of geomorphological advice will reflect perception of costs and benefits. For example, would the cost of not building in a possibly unstable area exceed, in the foreseeable financial future, the likely outlays on repairs, remedial work and compensation? Each local, state or national government will have to consider the degree of risk, the acceptable levels of risk and the likely effect of doing nothing about a potential problem. However, public and political pressure, may sway opinion and influence changes to building regulations, planning procedures and development control. The way this works after disaster has already been described, but all too often, action in one area leads to a spatial shift of problems to another locality. These off-site consequences are often neglected. They are part of the dynamics of the urban system, the physical response to technical change. Unless attitudes change, these off-site, downstream, consequential problems are likely to increase. Environmental impact assessment of flood control works, waste dumping and major construction projects ought to reveal the consequences of urban activities, but suburban growth, with channelization of minor streams, is seldom investigated in this way. The cumulative impacts of land cover change and resource use are so often the cause of consequential geomorphological problems.

Urban policy must not view geomorphology in terms of physical limitations, but as part of the changing dynamics of the urban system. In the old British cities, the clearance of former industrial land, the cessation of groundwater abstraction for industry and the replacement of paved areas by vegetation must alter infiltration rates and groundwater levels. Rising groundwater tables may affect foundations, drainage of slopes and rainfall–runoff relationships. Economic change and technological change affect the buildings, maintenance and land use of the city, which in turn influence the work of geomorphic processes. As the socio-economic character of the city alters, so there is a geomorphic response, with which urban policy ought to be designed to cope.

Policy has not only to cope with socio-economic change, but also with environmental change external to urban areas. Such global changes as the development of the Antarctic ozone hole and the increase of global atmospheric $CO_2$ concentrations could lead to a rise in temperature of 2 °C by the mid-21st century. As the temperature increase would be of the order of 5 °C over Antarctica, this could provoke the disintegration of the Antarctic ice-sheet with a concomitant sea-level rise of 4 to 6 m within a 100 years or so (Chen and Drake, 1986). Such a rate of sea-level rise should be compared with the present rate of rise of 4 m in 1000 years common on the east coast of the United States, and 6 m in 1000 years, largely as a result of subsidence, at Galveston, Texas (Komar and Holman, 1986). Urban policy and engineering technology should be able to cope, at least at the slower rate of rise, but recognition of the possible problem and its consequences must come early enough.

Sea-level rise, subsidence, flooding, slope instability, river channel change often come together in large coastal urban areas. Urban policy needs to examine the mutiple environmental and earth science problems which beset major settlements. A city like Tokyo requires constant reassessment of flood hazards and drainage problems as a result of subsidence, land reclamation and the occurrence of tropical cyclones. Geomorphology is part of this critical, ongoing appraisal of urban environmental conditions.

Clearly, there is a place for geomorphology in urban policy. That place is seldom clearly expressed, partly because geomorphology is rarely identified as a specific area of expertise and often is misused between ecological, engineering and geological advice. The general public ought to have a better appreciation of earth surface dynamics. Much could be done through schools and the media to have more geomorphology of the appropriate kind, in geography and science lessons and in natural history and science programmes.

Geomorphologists should play a bigger part in the education of the planners and engineers who advise local and national government decision makers. However, they must also be prepared to comment on local issues in which geomorphological criteria play a significant role and argue for local, and national policies that are geomorphologically sound. The majority of the world's people will live in urban areas by the year 2000. Urban geomorphology must not remain a side interest of the profession, but a task which must be tackled and promoted for the good of all.

## REFERENCES

Blair, M. L. and W. E. Spangle (1979). Seismic safety and land-use planning—selected examples from California, *US Geological Survey Professional Paper*, **941-B**, 82pp.

Bracegirdle, C. (1971). *The Dark River*, Sherratt, Altrincham. 246pp.

Brown, R. J. E. and G. H. Johnston (1964). Permafrost and related engineering problems, *Endeavour*, **23** (May), 66–72.

Butler, C. (Chairman) (1987). *Report of the Study Group into Geological Surveying*, London: Advisory Board for the Research Councils and Natural Environment Research Council.

Carter, J. A., P. J. Cresswell, P. T. Moore and J. R. Woolnough (1984). Design and development of the flood storage and amenity reservoirs of the Rother Valley Country Park, *Journal of the Institution of Water Engineers and Scientists*, **38**, 403–23.

Chen, C-T. A. and E. T. Drake (1986). Carbon dioxide increase in the atmosphere and oceans and possible effects on climate, *Annual Review of Earth and Planetary Sciences*, **14**, 201–35.

Cooke, R. U. (1984). *Geomorphological Hazards in Los Angeles* (The London Research Series in Geography 7), Allen & Unwin, London. 206pp.

Cooke, R. U., D. Brunsden, J. C. Doornkamp and D. K. C. Jones (1982). *Urban Geomorphology in Drylands*, Oxford University Press, Oxford.

Dewan Bandaraya (1981). *Kuala Lumpur Draft Structure Plan*, City Council, Kuala Lumpur.

Douglas, I. (1978). The impact of urbanisation on fluvial geomorphology in the humid tropics, *Geo-Eco-Trop*, **2**, 229–42.

Douglas, I. (1979). Flooding in Australia: a review. In R. L. Heathcote and B. G. Thom (eds) *Natural Hazards in Australia*, Australian Academy of Science, Canberra. pp. 143–63.

Douglas, I. (1985a). Geomorphology and urban development in the Manchester area. In R. H. Johnson (ed.) *The Geomorphology of North-West England*, pp. 337–52. Manchester University Press, Manchester.

Douglas, I. (1985b). Urban sedimentology, *Progress in Physical Geography*, **9**, 254–80.

Engels, F. (1845). *The Condition of the Working Class in England* (1971 edition, translated and edited by W. O. Henderson and W. H. Chaloner), Blackwell, Oxford.

Gupta, A. (1987). Urban geomorphology in the humid tropics: the Singapore case. In V. Gardiner (ed.) *International Geomorphology Part I*, pp. 303–17. Wiley, Chichester.

Hannam, I. D. and R. W. Hicks (1980). Soil conservation and urban land use planning, *Journal of Soil Conservation New South Wales*, **36**, 134–45.

Hannam, I. D., J. R. Davis and K. D. Cocks (1986). Implementing total catchment management strategies, *Journal of Soil Conservation New South Wales*, **42**, 80–82.

Hannam, I. D., K. A. Emery and B. W. Murphy (1978). *Urban Capability Study Bathurst*, New Wales Soil Conservation Service, Sydney.

Harrison, C. and J. R. Petch (1985). Ground movements in parts of Salford and Bury, Greater Manchester—aspects of urban geomorphology. In R. H. Johnston (ed.) *The Geomorphology of North-West England*, pp. 337–52. Manchester University Press, Manchester.

Helley, E. J., K. R. Lajoie, W. E. Spangle and M. L. Blair (1979). Flatland deposits of the San Francisco Bay Region, California—their geology and engineering properties, and their importance to comprehensive planning, *US Geological Survey Professional Paper 943*, 88pp.

Hough, M. (1984). *City Form and Natural Process*, Croom Helm, London. 281pp.

Johnston, D. (1986). Urbanisation and total catchment management, *Journal of Soil Conservation New South Wales*, **42**, 22–4.

Jones, A. (1970). Landslides of Rio de Janeiro and the Serra das Arras escarpment, Brazil, *US Geological Survey Professional Paper 697*, 42pp.

Komar, P. D. and R. A. Holman (1986). Coastal processes and the development of shoreline erosion, *Annual Review of Earth and Planetary Sciences*, **14**, 237–65.

Kremer, R. H. J. (1985). Management of groundwater: a daily job in our older cities, *Water in Urban Areas. TNO Committee on Hydrological Research Proceedings and Information*, **33**, 233–50.

Leggett, R. F. (1973). *Cities and Geology*, McGraw-Hill, New York.

Leigh, C. H. (1982). Urban development and soil erosion in Kuala Lumpur, *Journal of Environmental Management*, **15**, 33–45.

Lewis, K. V., P. A. Cassell and T. J. Fricke (eds) (1975). *Urban Drainage Design Standards and Procedures for Peninsular Malaysia*, Publications Unit Ministry of Agriculture, Kuala Lumpur. 279pp.

Massey, S. (1976). *A History of Stretford*, Sherratt, Altrincham. 328pp.

Mousinho de Meis, R. and da Silva. R. (1968). Movements de masse recents a Rio de Janeiro: une étude de géomorphologie dynamique, *Revue de Géomorphologie dynamique*, **18**, 145–51.

Noonan, G. A. (1986). An operation flood warning system, *Journal of the Institution of Water Engineers and Scientists*, **40**, 437–53.

Pearson, S. (1947). Potamology, or this river business, *Institution of Civil Engineers North Western Association Session 1946–47*, 16pp.

Pokrajcic, B. (1976). Hydric epidemics in karst areas of Yugoslavia, caused by spring water contaminations. In V. Yevjevich (ed.) *Karst Hydrology and Water Resources Vol II*, Fort Collins: Water Resources Publications, pp. 703–18.

Quinlan, J. F. (1986). Legal aspects of sinkhole development and flooding in karst terranes: 1 Review and Synthesis, *Environmental Geology and Water Sciences*, **8**, 41–61.

Rau, J. L. and P. Nutalaya (1982). Geomorphology and land subsidence in Bangkok, Thailand. In R. L. Craig and J. C. Craft (eds) *Applied Geomorphology* (Binghamton Symposia in Geomorphology International Series 11), pp. 181–201, Allen & Unwin, London.

Robinson, G. D. and A. M. Speiker (1978). Nature to be commanded . . . Earth-science maps applied to land and water management, *US Geological Survey Professional Paper, 950*, 95pp.

Salomone, W. G. (1986a). The applicability of the Florida Mandatory Endorsement for sinkhole collapse coverage. Part I. Legal aspects, *Environmental Geology and Water Sciences*, **8**, 63–71.

Salome, W. G. (1986b). The applicability of the Florida Mandatory Endorsement for sinkhole collapse coverage. Part II. Case history: foundation settlement of a residential structure—was it a sinkhole? *Environmental Geology and Water Sciences*, **8**, 73–6.

Sheedy, K. A., W. M., Lewis, A. Thomas and W. F. Beers (1982). Land use in carbonate terrain: problems and case study solutions. In R. L. Craig and J. L. Craft (eds) *Applied Geomorphology* (Binghamton Symposia in Geomorphology International Series 11), pp. 202–13. Allen & Unwin, London.

Singapore Drainage Department (1978). *Code of Practice on Surface Water Drainage*, Singapore: Ministry of Environment.

So, C. L. (1986). Geomorphic constraints on urban development: the Hong Kong experience. *Landplan II Role of Geology in Planning and Development of Urban Centres in South-east Asia, AGID Report Series 12*, pp. 59–74.

Tan, B. K. (1986). Geological and geotechnical problems of urban centres in Malaysia, *Landplan II: Association of Geoscientists for International Development, Report Series*, **12**, 10–14.

Tregay, R. and D. Moffat (1980). An ecological approach to landscape design and management in Oakwood, Warrington, *Landscape Design*, **132**, 33–36.

Turk, L. J. (1976). Predicting the environmental impact of urban development in a karst area. In V. Yevjevich (ed.) *Karst Hydrology and Water Resources, Vol II*, Water Resources Publications, Fort Collins.

Yap, K. G. (1985). *Prevention and Control of Floods in Singapore*, Paper presented to the Inter-Faculty Seminar on Rainfall, Floods and Planning in Singapore, November 1985.

Ziony, J. I. (ed.) (1985). Evaluating earthquake hazards in the Los Angeles Region—an earth science perspective, *US Geological Survey, Professional Paper 1360*, 505pp.

Geomorphology in Environmental Planning
Edited by J. M. Hooke



# 6 Heavy Metal Contamination in Soils of Tyneside: A Geographically-based Assessment of Environmental Quality in an Urban Area

**RICHARD ASPINALL**
*The Macaulay Land Use Research Institute, Craigiebuckler, Aberdeen*

**MARK MACKLIN and STAN OPENSHAW**
*Department of Geography, University of Newcastle, Newcastle Upon Tyne*

## INTRODUCTION

Recently there has been an increasing awareness among local authorities, central government and the general public of environmental contamination by heavy metals, and the potential danger they pose to human health and ecosystems. This growing concern has been reflected perhaps most notably by various reports on the effects of lead (e.g. DHSS, 1980), by the Royal Commission on Environmental Pollution in its ninth report (1983) *Lead in the Environment*, and in a series of pollution papers focusing on heavy metals such as cadmium (DoE, 1980) and mercury (DoE, 1976). Increasing efforts are seemingly now being made to reduce the amount of heavy metals being discharged into the environment, through stricter controls, and, in particular instances, by more effective monitoring of transport pathways and storage mechanisms of trace metals.

A wider appreciation of the nature of the problem has been hampered by the general absence in Britain (and in many other countries also) of systematic data on the distribution of metal contaminants in the environment mapped at an appropriately fine level of geographical detail. Nowhere has this been more true than in the evaluation of metal pollution in soils, especially in urban areas. Difficulties arise most often here because of problems in ensuring consistent sampling procedures, distinguishing natural from historic and contemporary anthropogenic effects, and also through strong local variability in soil metal concentrations. In addition, interpretation of soil metal patterns and levels is frequently problematic because the long industrial history of Britain, with a continuous environmental overprinting by metal contaminants, makes it difficult to establish unequivocally how far present metal levels in urban soils can be attributed to current or past industrial activities. Raised levels of heavy metals in soils is an issue of considerable importance wherever they impinge directly on human health. This may occur when food is grown on contaminated land, or when metal-polluted soil is inadvertently ingested by young children. In the latter case there is growing epidemiological evidence for a link between Pica-related high blood lead levels and impaired IQ (DHSS, 1980). Soil pollution also has major planning implications especially where new industry is sited in a historically metal contaminated area. For although atmospheric metal emissions may comply with those set by statute, a comparatively small increase of heavy metals in soils around a works could result in unacceptably high metal burden for local residents.

There are comparatively few studies (though see JURUE (1982) report for Walsall and Parry, Johnson and Bell's (1981) study of Merseyside) which have assessed soil metal levels and their geographical variation at anything approaching a reasonable level of detail within a single urban area in Britain. This scarcity of studies is surprising when one considers the long residence time of metal pollutants in soils (over timespans to be measured in centuries and millennia) and because information on trace metal contamination is an essential prerequisite before a benchmark or backcloth can be established against which polluted areas can be compared, and links made between metals and public health. In the light of this apparent neglect, a programme of soil sampling and heavy metal analysis was initiated in Tyneside. With a long history of industrial development, Tyneside provides a geographically distinct area for a preliminary soil metal survey, and also an opportunity to develop and apply sampling and analytical methodologies that may be appropriate for investigating soil pollution in other urban environments.

## AIMS OF SURVEY

There were four principle aims of this survey:

(1) To establish, initially at a 1 $km^2$ scale and then on a ward basis, concentrations of heavy metals in the soils of Tyneside.
(2) To distinguish between background soil metal levels and anomalous concentrations using a series of statistically based contamination 'thresholds'.
(3) To identify, from these thresholds, metal contaminated 'black spots' and to assess the extent to which these are historical legacies or attributable to contemporary human activities.

(4) To estimate, for the first time in a major British urban area (to the knowledge of the authors), the number of people living in metal contaminated areas. This has important implications for the future siting of heavy metal producing industries, for the redevelopment of certain areas, and is an essential prerequisite for subsequent epidemiological studies.

## STUDY AREA

Tyneside consists of four Metropolitan Districts, Newcastle, Gateshead, North and South Tyneside which have a combined population approaching one million. The area was among the cradles of the industrial revolution and since the turn of the eighteenth century has been a centre for shipbuilding, engineering, the chemical industry, non-ferrous and ferrous metal processing. It is arguably one of the most intensely industrialized and highly populated regions of the UK.

Most of Tyneside is underlain by productive coal measures of Upper Carboniferous age, but in the south-west of the coalfield these are concealed by Permian sands and limestones. Dolerite dykes of two ages, Late Carboniferous and Tertiary, cut the Upper Carboniferous rocks of the area. Quaternary drift thickness are highly variable; the drift is generally thin ($<2$ m) on land above 76 m OD, but within the buried, bedrock-cut valley systems of the Tyne, Derwent, and Team superficial deposits usually exceed 20 m (Dearman *et al.*, 1979). Newcastle city has a mean annual rainfall of 667.6 mm based on measurements by the Department of Geography, University of Newcastle, over the period 1952–85. Prevailing wind directions are from the west, south-west and north-west.

## SOIL SAMPLING METHODS AND HEAVY METAL ANALYSIS

Four hundred and twelve soil samples were collected from 320 pre-selected 1 $km^2$ grid squares. Sampling points within these were located to the nearest 100 m using random numbers. As far as possible samples, taken from the top 15 cm of the soil and weighing between 0.5 and 1.0 kg, were collected from relatively undisturbed grassy areas in public open spaces.

In the laboratory soils were air-dried for 48 hours, lightly ground using a pestle and mortar, and passed through a 2 mm mesh sieve. 'Total' heavy metal contents quoted in this chapter were determined as the amounts of these elements brought into solution by digesting 5 g of soil in 25 per cent nitric acid. 'Plant-available' metals were estimated in solutions extracted using 0.5 m acetic acid, after shaking 5 g of soil in 200 ml of solvent for one hour. Metals extracted from soil using acetic acid are considered to be present in a chemically 'active' form more readily available for uptake by plants. This chemically 'active' fraction is also included in the 'total' metal content of a soil, though more vigorous nitric acid digestion additionally releases metals present in a relatively chemically inert form, that may only constitute a danger when contaminated material is directly ingested by animal or man. Metal concentrations were measured by atomic absorption spectrophotometry (air/acetylene flame) using a Pye Unicam SP9 instrument with computer. All soil metal concentrations are the means of three assays.

## ESTIMATION OF BACKGROUND METAL CONCENTRATIONS IN TYNESIDE SOILS: SETTING CONTAMINATION THRESHOLDS

Metal contaminated and uncontaminated soils in Tyneside were distinguished using a graphical estimation method as outlined by Davies (1983). Percentage

**Table 6.1** (a) Highest probable values ('thresholds') for 'total' and 'plant available' lead, zinc and cadmium in uncontaminated Tyneside soils

| | Metal | Metal concentration (mg/kg) |
|---|---|---|
| 'Total' | Cadmium | 1 |
| | Lead | 80 |
| | Zinc | 345 |
| 'Plant available' | Lead | 14 |
| | Zinc | 11 |

(b) Some statistically-based threshold values for 'total' lead in soils

| Author | Study area | Threshold (mg/kg) |
|---|---|---|
| Archer (1980) | Agricultural soils in England and Wales | 108 |
| Bradley and Cox (1986) | Hamps and Manifold valleys, North Staffordshire | 53 |
| Davies (1983) | Ceredigion | 116 |
| | Tamar Valley | 107 |
| | Halkyn Mountain | 90 |
| | North Somerset | 110 |
| Lewin, Bradley and Macklin (1983) | Ystwyth Valley, mid-Wales | 125 |
| Macklin, Bradley and Hunt (1985) | Lox Yeo Valley, Somerset | 67 |
| This study | Tyneside | 80 |

(c) ICRCL tentative guidelines on acceptable levels of lead in soils for four types of redevelopment

| Vegetable gardens and allotments | Other gardens | Amenity land | Public open space |
|---|---|---|---|
| (metal concentration mg/kg) | | | |
| 550 | 550 | 1500 | 2000 |

cumulative frequency distributions of the $\log_{10}$ metal content for all soils were plotted on probability graph paper and a lower, linear portion of each curve was identified, separated, replotted and interpreted as a lognormal population derived from non-contaminated soils. From this, geometric means and deviations were calculated allowing the highest probable (threshold) metal value for an uncontaminated soil to be estimated. The results of these calculations are given in Table 6.1, and, for comparison, other statistically based threshold values as well as the ICRCL (1980) guideline for acceptable levels of lead in soils.

Tyneside soils have a rather low uncontaminated/contaminated soil lead threshold. This may in part reflect that other soil lead thresholds, excluding those of Archer (1980), are estimates from mineralized areas of England and Wales where metal mining has caused gross metal contamination and 'normal' soil metal concentrations are probably somewhat higher. It is of some interest to note, however, that even the most stringent guidelines of acceptable lead concentrations in soils exceed in every study the highest concentration of lead expected in an uncontaminated soil by a factor of between four and eight (Table 6.1).

## MAPPING SOIL CONTAMINATION IN TYNESIDE

The soil data were analysed as point data (using 100 m grid references) and by aggregation to 1981 census wards. The latter process used a point in polygon procedure from the GAG library and digital boundary representations of the census wards. This aggregation to wards is important for two reasons:

(1) It allows chloropleth maps to be drawn.
(2) It is possible to link soil metal results with socio-economic data which are reported by wards from the 1981 census.

Both processes, however, do necessitate a degree of geographic generalization. Additional interpretive problems arise where wards contain a variable number of sample points. The resulting mean heavy metal concentrations calculated for each ward may not therefore, be completely representative of all points within that ward but this process does at least allow general comparisons to be made and maps to be drawn.

The class intervals used for the chloropleth maps of 'total' and 'plant available' cadmium, lead and zinc concentrations in Tyneside are based on a selection of published guidelines of acceptable soil metal concentrations (Berrow and Burridge, 1977; ICRCL, 1980; Archer, 1980) and our own statistically based estimates of threshold values (Table 6.1). However, for 'total' soil lead concentrations an intermediate class interval of 150–550 mg/kg was included in the light of the Royal Commission on Environmental Pollution ninth report

(1983) statement that, 'soil lead levels above 150 $\mu g\ g^{-1}$ are most likely to have arisen from contamination' (para 2.16, P19).

In order to evaluate the extent of soil metal contamination on Tyneside the percentage of soils sampled exceeding our own thresholds and published guidelines were calculated, as was the proportion of the population living in wards in which soil metal concentrations exceeded respective class intervals (Table 6.2). These latter percentages illustrate in a general way how many people in Tyneside are living in areas in which soils are contaminated to a greater or lesser extent by heavy metals.

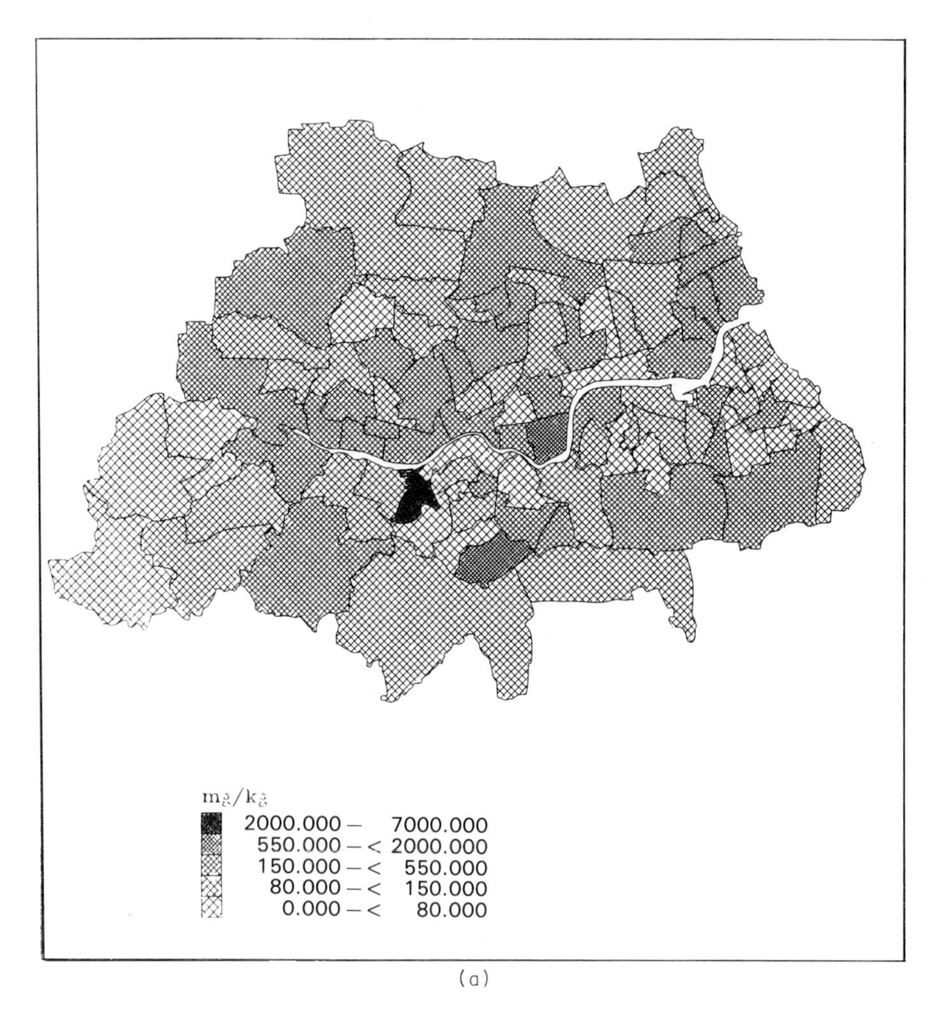


**Figure 6.1** (a) Total lead in soils in Tyneside census wards. (b) Plant-available lead in soils in Tyneside census wards

## SPATIAL VARIABILITY IN METAL CONTAMINATION IN TYNESIDE SOILS

The spatial variation in soil metal concentrations for both total and plant available metals are shown in a series of chloropleth maps of Tyneside census wards (Figures 6.1–6.3). Soils whose 'total' lead concentrations are greater than 80 mg/kg constitute 69.1 per cent of the samples analysed in Tyneside and are found in most parts of the city. Higher threshold values show more restricted distributions; concentrations of between 150 and 550 mg/kg (35.4 per cent) are found in soils within central Newcastle and along the River Tyne. The areas from Wallsend to Newburn and from Blaydon to Ryton are in this class,

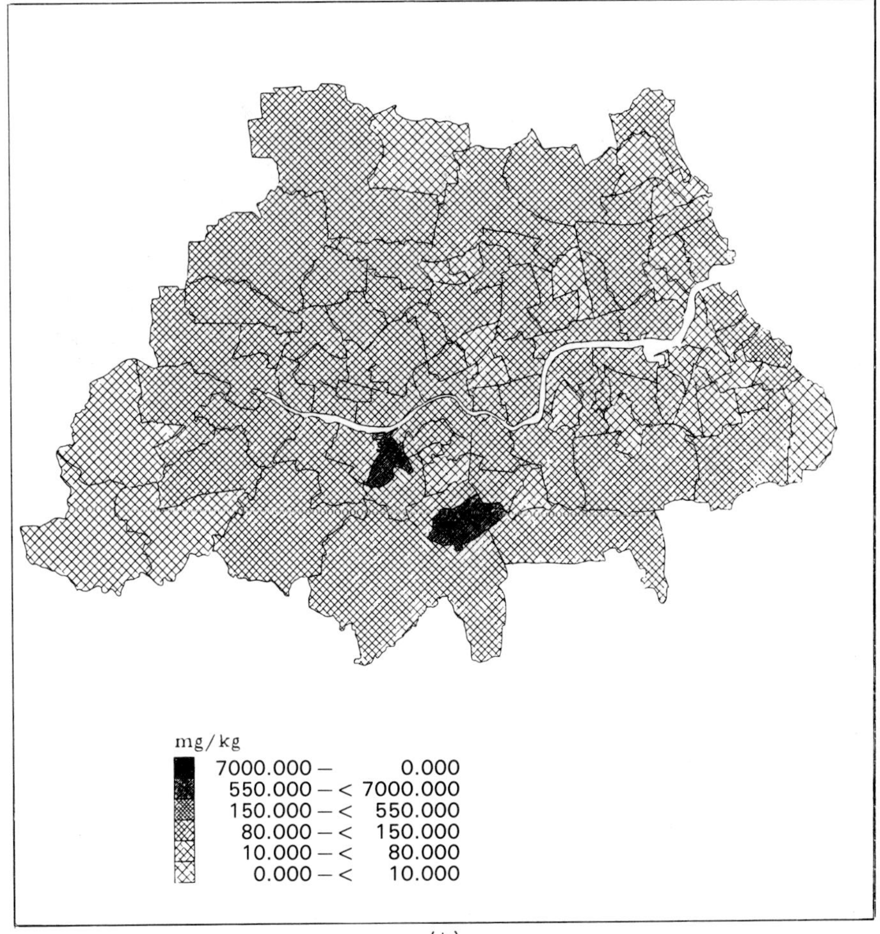


(b)

as are Boldon, Cleadon, Fenham, Kenton and the coast between Cullercoats and Whitley Bay. Four areas (6.5 per cent of soils sampled in Tyneside) have concentrations over 550 mg/kg; along the north of the Tyne around Elswick and Walker, and south of the river in the Harlow Green—Wrekenton and Team Valley areas. The highest mean concentration by ward (2000 mg/kg) is in the Team Valley area.

'Available' lead occurs at low concentrations across the whole of the Tyneside area and was in excess of 10 mg/kg in 65 per cent of soil samples. Concentrations in excess of 50 mg/kg were more restricted with clusters in the Jarrow, Byker, Heaton, and Woolsington areas, and also around Wrekenton and South Shields.

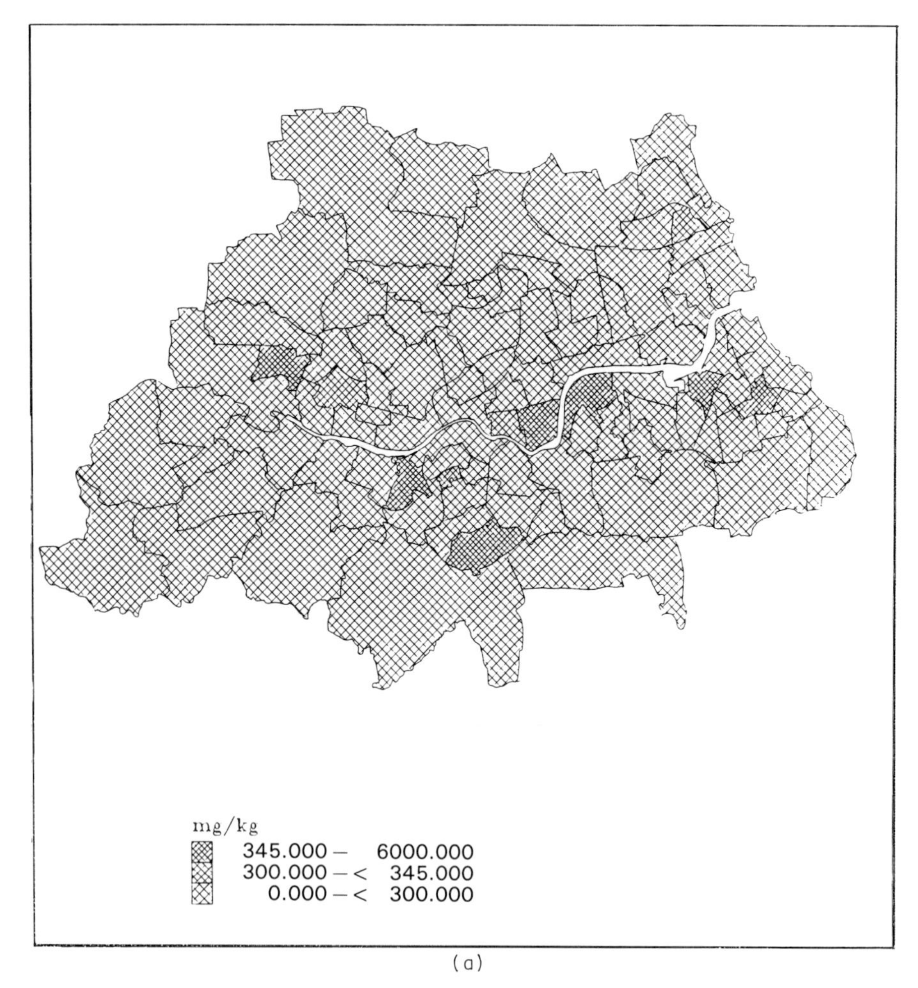


(a)

**Figure 6.2** (a) Total zinc in soils in Tyneside census wards. (b) Plant-available zinc in soils in Tyneside census wards

Highest available lead concentrations in soil (80 mg/kg) occur in the Boldon, Cleadon, Woolsington, Byker and Wrekenton wards.

Zinc concentrations in excess of 300 mg/kg occur in 14.4 per cent of samples analysed, these being distributed widely throughout Tyneside. The ward map (Figure 6.2) clearly shows high values to be clustered around the River Tyne, especially in the shipyard and engineering areas of Jarrow, Hebburn, Walker and near Derwenthaugh, these having been sites of such industries since before 1760.

Concentrations of total cadmium greater than 1 mg/kg occur in 47.4 per cent of soil samples, and these are, as for lead and zinc, widely distributed throughout Tyneside (Figure 6.3). Concentrations in excess of the ICRCL (1980) threshold of 3 mg/kg occur in 6.1 per cent of the samples and these have a more restricted

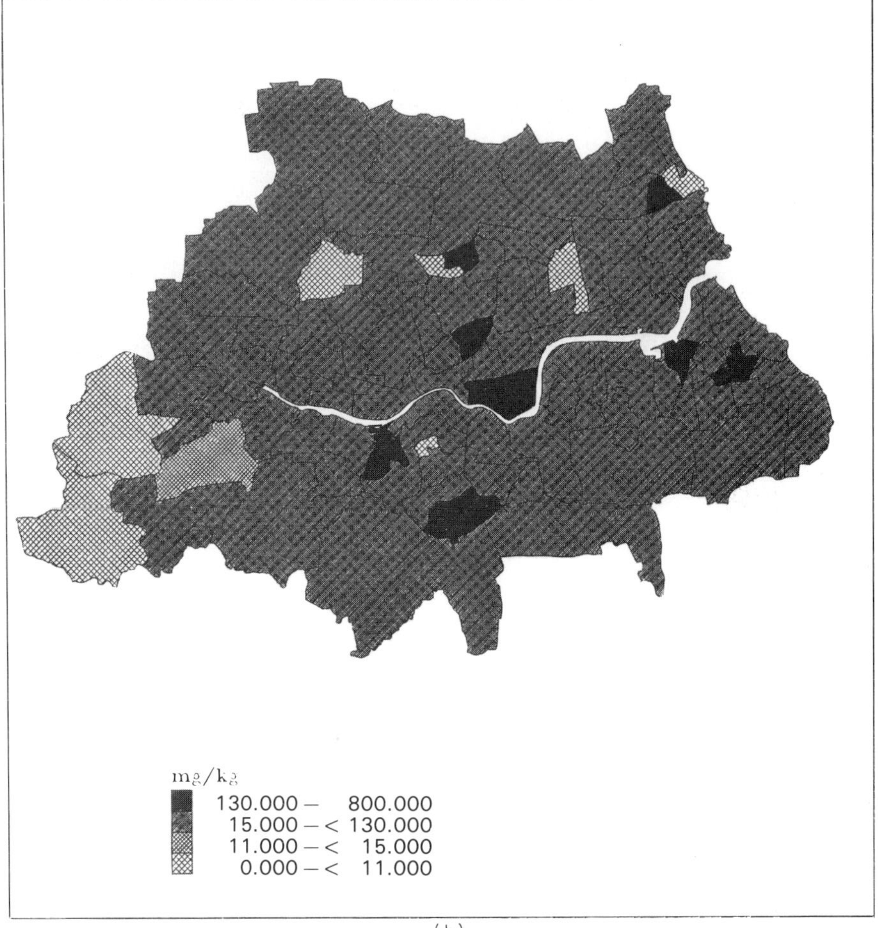


(b)

distribution, being found in three areas. First, high values were recorded in the area around Byermoor, close to Winlaton Mill; second, in areas to the south of the Tyne and towards the coast around Jarrow and South Shields; and third, in a large number of samples from north-east Tyneside towards Newcastle airport and Ponteland.

## POPULATION DISTRIBUTION IN RELATION TO CONTAMINATED SOILS IN TYNESIDE

The proportion of the Tyneside population living in wards where soil metal concentrations exceed the various threshold levels are shown in Table 6.2, these

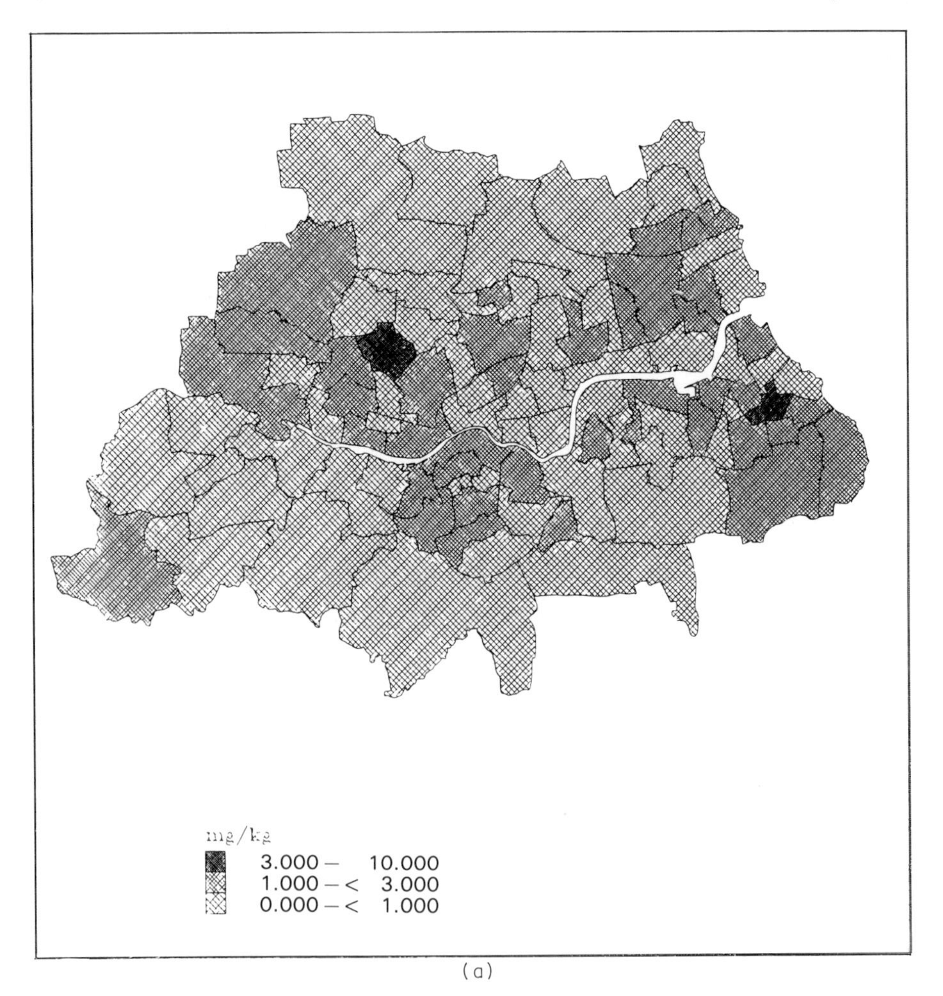


(a)

**Figure 6.3** (a) Total cadmium in soils in Tyneside census wards. (b) Plant-available cadmium in soils in Tyneside census wards

data revealing several important points. First, the lower the threshold, the greater the proportion of the population living in contaminated areas. Second, at the most stringent levels for the 'total' form of the various metals, the proportion of the population in contaminated areas is greatest for lead (84.3 per cent) and lowest for zinc (11.8 per cent), cadmium being intermediate (47.4 per cent). This order is different for 'plant-available' forms of metals; zinc contaminated areas containing the highest proportion of the population (97.1 per cent) and lead contaminated areas 80.3 per cent. Additionally, a greater proportion of the population is exposed to elevated concentrations of plant-available forms of metals than total metals. There is the possibility therefore that vegetables grown in allotments and gardens in Tyneside could represent a significant pathway by which metals are transferred from contaminated soil to the local populace. This raises important questions on the definition of 'acceptable' levels for both total and plant-available metals

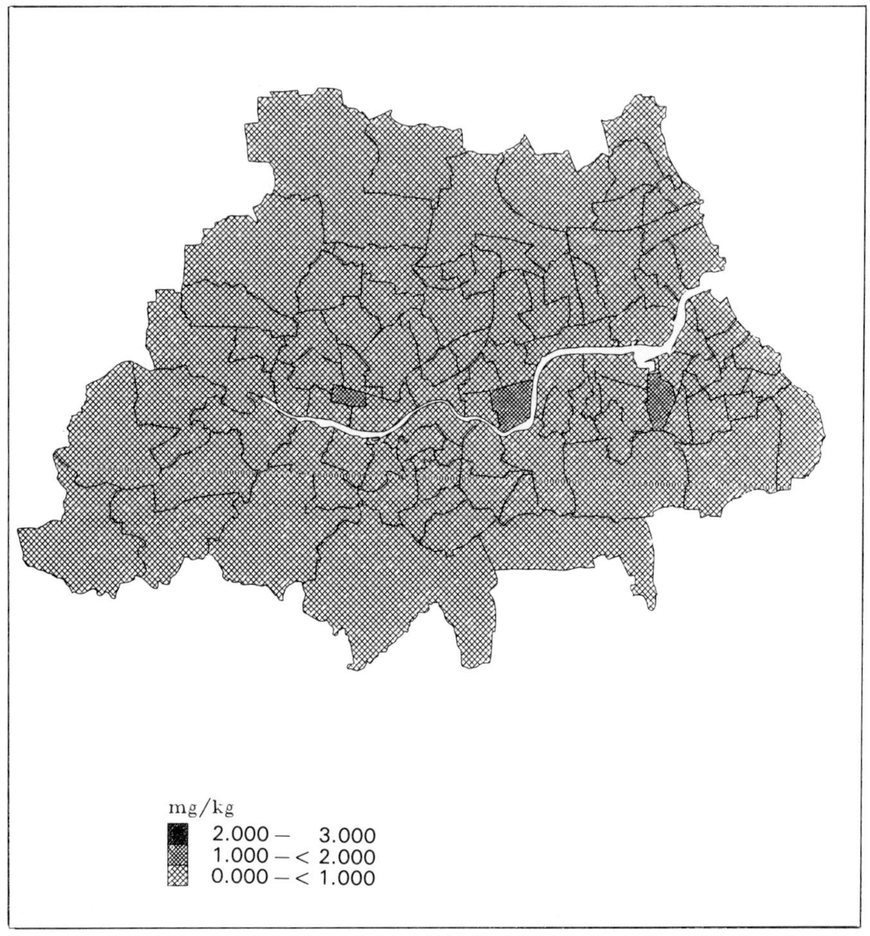


(b)

**Table 6.2** Percentage of Tyneside population living in wards where measured values exceed thresholds, and percentage of soils sampled exceeding thresholds

| Metal | Class Intervals (mg/kg) | Source of threshold value | Percentage of Tyneside soils sampled exceeding class interval | Percentage of Tyneside population living in contaminated wards |
|---|---|---|---|---|
| (a) 'Total metals | | | | |
| Cadmium | 1 | This study | 47.4 | 42.4 |
| | 3 | ICRCL (1980) | 6.1 | 2.2 |
| Lead | 80 | This study | 69.1 | 84.3 |
| | 150 | Royal Commission 9th Report (1983) | 35.4 | 44.0 |
| | 550 | ICRCL (1980) | 6.5 | 4.8 |
| | 2000 | ICRCL (1980) | 1.3 | 0.7 |
| Zinc | 300 | ICRCL (1980) | 14.4 | 11.8 |
| | 345 | This study | 11.2 | 9.1 |
| (b) 'Available' metals | | | | |
| Cadmium | 2 | Berrow and Burridge (1977) | 0.3 | 0.0 |
| Lead | 10 | Berrow and Burridge (1977) | 65.0 | 80.3 |
| | 14 | This study | 49.0 | 58.0 |
| Zinc | 11 | This study | 78.8 | 97.1 |
| | 15 | Berrow and Burridge (1977) | 68.7 | 95.7 |
| | 130 | ICRCL (1980) | 6.7 | 9.4 |

currently used, a minor proportion of the Tyneside population living in areas considered contaminated by the ICRCL levels, a much higher proportion living in areas of actual contamination. Questions on the definition of acceptable levels are especially relevant for total lead and cadmium and for plant-available forms of lead and zinc. Finally, the importance of plant-available forms of metals as contaminants of areas now heavily populated raises questions concerning methods of measuring appropriate forms of metals in soils and linking these data with metal transfer pathways through soil–plant–animal systems and human health.

## SOURCES OF METAL CONTAMINATION IN TYNESIDE SOILS

Present-day levels of heavy metals in soils of Tyneside and their distribution can be attributed to a number of sources that have overprinted 'natural'

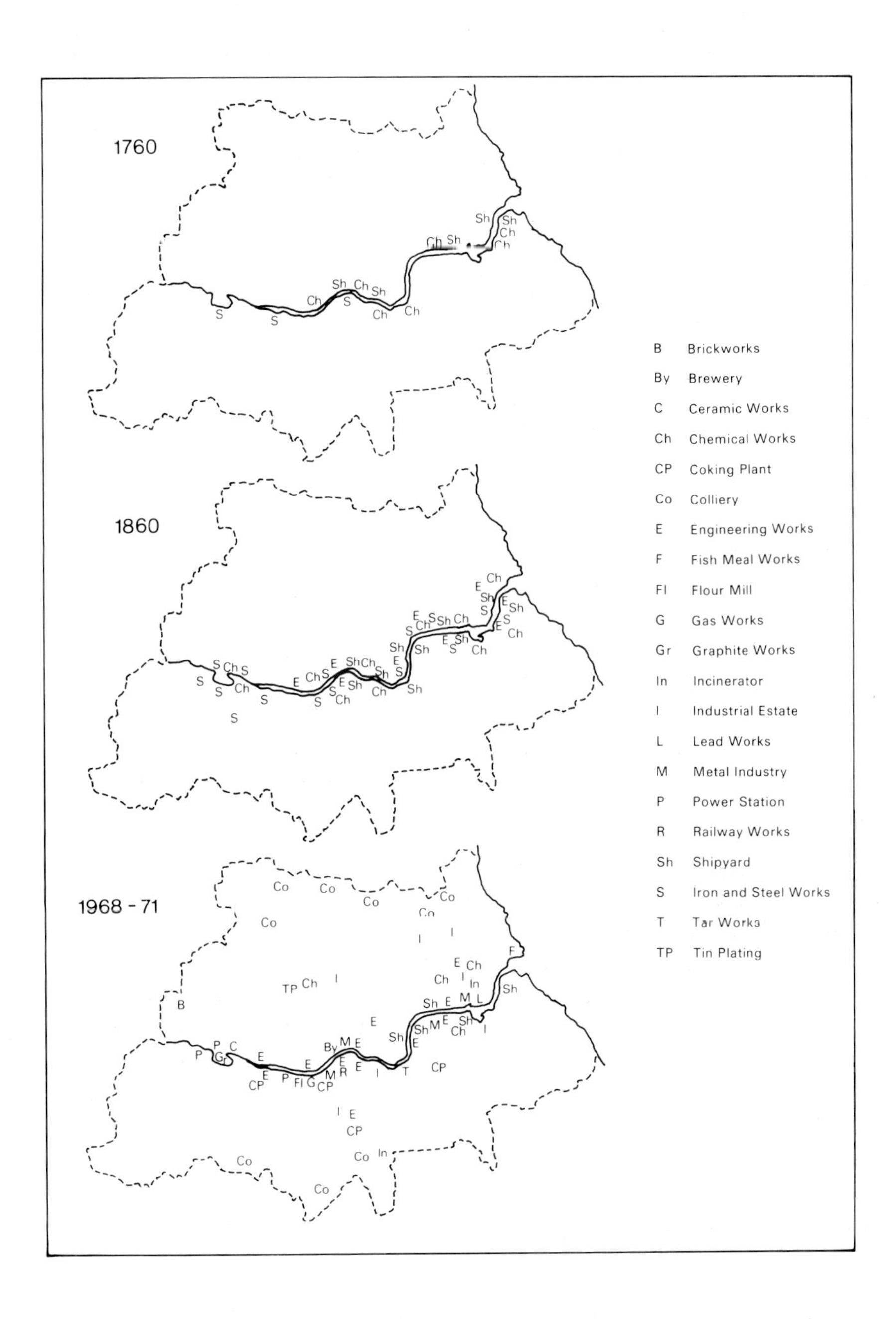


**Figure 6.4** Distribution of industry on Tyneside: 1760, 1860, 1968–71

geologically-controlled metal concentrations and include both past and present industry, and vehicle emissions. The distribution of soils contaminated by lead and zinc appears to be closely associated with shipbuilding, heavy engineering and chemical industries which have been concentrated along parts of the River Tyne waterfront from the beginning of the eighteenth century (Figure 6.4). Indeed, industrial development since that time has remained focused along the river, and metal concentrations in soils show a high degree of contamination along the entire length of the river as a result of the long history of pollution by industry. The road network too, shows some association with the distribution of lead in soils, this being most notable in the coastal areas of North Tyneside between Whitley Bay and Tynemouth and in Newcastle city centre, although this area was also the location of a number of lead works during the nineteenth century. Recent industrial development in the Team valley and an incinerator at Wrekenton have resulted in high concentrations of metals in local soils.

The map for cadmium in soils discloses high concentrations towards the airport and Ponteland where natural cadmium-rich soils have been derived from Carboniferous black shales and also cadmium contaminated soils, notably to the south of the River Tyne around Jarrow and South Shields, and near to Winlaton Mill at Byermoor, in the older heavily industrialized areas of Tyneside.

## IMPLICATIONS OF HEAVY METAL CONTAMINATION IN TYNESIDE SOILS AND DIRECTIONS FOR FUTURE RESEARCH

A number of conclusions may be drawn from this study, that throw some light on the magnitude of soil contamination by heavy metals and accepted guidelines, and to the spatial extent and causes of metal contamination on Tyneside.

First of all, it is clear that over two-thirds of soils sampled in Tyneside contained concentrations of total lead in excess of our statistically based contamination threshold of 80 mg/kg. If, however, the most stringent ICRCL (1980) guideline for acceptable total lead is used only 6.5 per cent of soils exceed this level (Table 6.2). Clearly there is a disparity between the ICRCL definition of 'acceptable' levels of contamination and actual levels of contamination that occur in the Tyneside area. By comparison, relatively few samples are considered to be contaminated by total zinc (Table 6.2), and significantly there is a coincidence between our statistically based threshold of 345 mg/kg and the ICRCL acceptable level of 300 mg/kg.

Based on Berrow and Burridge's (1977) data for range of available metals typically found in soils, the percentage of soils contaminated with available lead is similar to that for total lead contamination However, a much greater proportion of soils are contaminated with plant-available zinc than total zinc, and very few soil samples show evidence of contamination with available or total cadmium.

As regards the pattern and causes of soil contamination on Tyneside by heavy metals, lead and zinc concentrations both show distinct geographical patterns that are clearly related to past and present industry. The areas with the greatest degree of contamination are associated with the historic heartland of industrial activity along the River Tyne, but there are also parts of Gateshead with high metal concentrations which are seemingly associated with incinerators and modern industrial operations. Lead concentrations in soils, however, also show some general relationship with the Tyneside road network confirming several previous studies in British cities that have established a relationship between the motor-car and lead levels in urban environment (Davies and Holmes, 1972). Finally cadmium shows a distribution which is principally controlled by local geology although there are areas to the south of the River Tyne where industry could be an important contributory factor.

The findings of this survey underline the pressing need for further and more detailed work on metal contamination in soils not only in Tyneside but also in other major conurbations in the UK. It is surprising that spatially detailed studies of heavy metal levels do not yet exist for the major industrial and urban areas of Great Britain and Europe given their well known adverse health effects. This reiterates Bullock's (1987) recent call for increased monitoring and the need to establish a more extensive data base of soil quality in the European Community. This study has examined, through focusing on a single British conurbation, metal contamination in soils at a comparatively 'coarse' intra-urban scale and it has become clear that there is a need to examine the contaminated 'hot-spots' on Tyneside in greater detail. At present our soil sampling network is too coarse, and the range of heavy metals analysed too few, to enable us to provide firm guidelines for planning and policy recommendations. Furthermore, techniques of chemical analysis currently used for measuring 'total' and 'plant-available' metal concentrations in soils are not really sufficiently sensitive to forewarn of possible health effects of high levels of potentially toxic metals in the environment. Future work should seek to ascertain the chemical forms of trace metals in soils which have major implications for understanding and predicting metal transfer within soil–plant–animal systems. The disparity between 'acceptable' and actual levels of contamination in Tyneside soils should also be investigated in other British cities and perhaps in agricultural areas also, special attention being paid to population distribution in these areas. Finally, though this study has drawn attention to the potential problems of metal contamination in an urban area, it should not lead us to overlook other equally toxic chemical components, including dioxins and PCBs, of which little is known of their concentration and spatial distribution in urban environments.

## Acknowledgements

The authors wish to extend special thanks to Mr Watts Stelling for his unerring technical support during this project. M.G.M. thanks N.E.R.C. for the provision of a Research Fellowship during the tenure of which this chapter was written.

## REFERENCES

Archer, F. C. (1980). Trace elements in soils in England and Wales. In *Inorganic Pollution and Agriculture*, Ref. Book 326, MAFF, HMSO, London. pp. 184–90.

Berrow, M. L. and J. C. Burridge (1977). Trace element levels in solids: effect of sewage sludge. In *Inorganic Pollution and Agriculture.* Ref. Book 326, MAFF, HMSO, London. pp. 159–83.

Bradley, S. B. and J. J. Cox (1986). Heavy metals in the Hamps and Manifold valleys, north Staffordshire, U.K.: distribution in floodplain soils, *The Science of the Total Environment*, **50**, 103–28.

Bullock, P. (1987). Soil protection—a need for a European programme. In H. Barth and P. L'Hermite (eds) *Scientific Basis for Soil Protection in the European Community*, Elsevier, pp. 581–7.

Davies, B. E. (1983). A graphical estimation of the normal lead content of some British soils, *Geoderma*, **29**, 67–75.

Davies, B. E. and P. L. Holmes (1972). Lead contamination of roadside soil and grass in Birmingham, England in relation to naturally occurring levels, *J. Agric. Soc. Camb.*, **79**, 479–84.

Dearman, W. R., M. S. Money, A. D. Strachan and A. Marsden (1979). A regional engineering geological map of the Tyne and Wear County, N.E. England, *Bull. of Int. Assoc. Eng. Geo.*, **19**, 5–15.

DHSS (1980). *Lead and Health: the Report of a DHSS Working Party on Lead in the Environment*, HMSO, London.

DoE (1976). Environmental mercury and man, *Pollution Paper No. 10*, HMSO, London.

DoE (1980). Cadmium in the environment and its significance to man, *Pollution Paper No. 17*, HMSO, London.

ICRCL (1980). *Tentative Guidelines for Acceptable Concentrations of Contaminants in Soil*, Interdepartmental Committee on the Redevelopment of Contaminated Land, Paper 38180, DoE.

JURUE (1982). Background levels of heavy metal soil contamination in Walsall. Joint Unit for Research on the Urban Environment, University of Aston in Birmingham.

Lewin, J., S. B. Bradley and M. G. Macklin (1983). Historical valley alluviation in mid-Wales, *Geol. Jour.*, **18**, 331–50.

Macklin, M. G., S. B. Bradley and C. O. Hunt (1985). Early mining in Britain: the stratigraphic implications of heavy metals in alluvial sediments. In N. R. J. Feiller, D. D. Gilbertson and N. G. A. Ralph (eds) *Palaeoenvironmental Investigations: Research Design, Methods and Data Analysis*, pp. 45–54, BAR International Series 258.

Parry, G. D. R., M. S. Johnson and R. M. Bell (1981). Trace metal surveys of soil as a component of strategic and local planning policy development, *Env. Poll.* (Series B) **2**, 97–107.

Royal Commission on Environmental Pollution (1983). *Lead in the Environment*. Ninth Report, HMSO, London.

# Slope Management

Geomorphology in Environmental Planning
Edited by J. M. Hooke



# 7 Slope Instability, Planning and Geomorphology in the United Kingdom

**DENYS BRUNSDEN**
*Department of Geography, King's College, London*

## INTRODUCTION

In 1978 applied geomorphology was defined as:

> the application of geomorphological techniques and analyses to the solution of a planning, environmental management, engineering or similar problem (Brunsden, Doornkamp and Jones, 1978).

Since then there have been rapid developments in the application of the subject particularly with respect to specific environment or engineering projects and it is now standard practice for geomorphological investigations to be incorporated in their reconnaissance and design stages (e.g. Doornkamp *et al.*, 1979; Cooke *et al.*, 1982; Jones, 1980; Brunsden, Doornkamp and Jones, 1979).

Far less attention has been given, however, to planning and decision-making aspects until the recent far-sighted operations of the Department of the Environment became influential through their Geological and Minerals Planning Research Programme contracts.

The purpose of this chapter, therefore, is to discuss the needs of the planner for earth science data. For illustrative purposes it will be restricted to discussions of slope instability but the conclusions do have much wider relevance. The chapter is intended as a personal, working geomorphologist's view on the public responsibilities of the subject.

## THE PLANNING PROCESS

Planning is defined here as 'that process of decision-making that aims to develop and conserve the land in the public interest'.

It is a system that provides a positive approach to development by the preparation of policy, plans and statements. It provides the development control function by making decisions with reference to specific development proposals; by reconciling conflicting claims on land use; by recycling land to new uses and by seeking to maintain or enhance environmental quality. Generally, each development proposal is treated on its merits and normally approved unless it can be shown that harm will be caused by the proposal. The process involves:

—The Department of the Environment, the Scottish Development Department and the Welsh Office who are responsible for overall land-use planning, policies and guidance.
—The County Councils for Structure Plans and Control of Mineral Extraction.
—The District and Local Councils for local plans and general control of specific developments.

These authorities act within a planning legislation framework provided by:

—the Town and Country Planning Act (1947) which was the first major legislation for the control of the development of land and which, together with amendments, additions and recent new legislation provide the basis for a refined, realistic and positive system of future development. In the context of this chapter, however, it must be noted that the Development Plans designed under this Act seldom identified 'risk zones' and most areas were allocated for development oblivious of slope instability risk. Flooding and subsidence were both recognized and provided for due to the provisions of other Acts.
—the Town and Country Planning Act (1971) and the local Government Planning and Land Act (1980) consolidated intervening amendments and especially the reformations of 1968, which produced the present system of 'structure' and 'local' plans. Again, however, these Acts, their policy plans and subsequent legislations do not have specific policies regarding landslips. Instead planning authorities appear to have an interest and responsibility for such events, implicitly rather than explicitly, through policies that only indirectly cover the subject (Flynn, 1985) in the following manner:

(1) Through policies aimed at maintaining landscape quality. This is particularly relevant where the occurrence of a landslide provides 'knock-on' effects that detract from landscape quality. For example, landslide remedial works may alter landscape through massive construction, earthworks or the creation of new borrow pits for aggregate or fill. These would be the subjects of planning control.
(2) The control of mineral extraction and subsequent restoration during which landslides might occur.

(3) In land reclamation programmes and disposal of waste (e.g. possible tailing pond failure).
(4) In the maintenance and prevention of obstruction of highways, under the Highway Act (1980).
(5) In the safety requirements of the Mines and Quarries (Tips) Act 1969 where there is a duty to ensure that such operations do not by reason of instability constitute a danger to members of the public.
(6) In the emergency procedures of the Local Government Act 1972 where there are immediate engineering, environmental and social responsibilities as well as the need to maintain access, communications and services in averting, handling or curing a disaster (Clarke, 1985).
(7) Through legal considerations such as the landowner's responsibility, resources of the Defendant, abatement of danger, knowledge of danger, or statutory duty to serve notice. Of primary importance, since the responsibility of an authority has never come before the Courts directly, is whether landsliding is a 'material consideration' to be taken into account by a planning authority and the question of 'Duty of Care' (Lambert, 1985).

It might also be noted that an Authority can be hampered considerably by the lack of legislation regarding powers of access. An Authority has no power to enter on land without the owner or occupier's consent whether or not there is a risk.

In view of this considerable uncertainty or lack of specific legislation it would be unreasonable to expect planners to have the necessary knowledge themselves to identify potentially unstable areas; they need expert advice (Payne, 1985). It is not the function of a local planning Authority to investigate the ground conditions of a site and probably they would not wish to become too heavily involved in this because to do so would tie up available financial resources that might be used more effectively elsewhere. Such costs are part of the capital costs of the development project itself.

Nevertheless, Authorities should, can and will take knowledge of landslides into account as an assumed material consideration in the determination of a planning application mainly because it helps the planning process to do so. It is widely recognized that consideration of actual or potential landsliding at subsequent stages, such as at the time of building regulation approval, is insufficient and too late to be of benefit to the developer. This is actually a fundamental point because the primary responsibility for deciding whether the land is physically suitable for a proposed development rests with the developer. This leaves the Authority free to seek additional information from the proposer or to assess the degree of investigation and remedial work to be applied, before itself coming to a view as to the likely consequences of the development.

The Authority then has a duty to warn a developer of the possibility of instability and may choose whether to give informal advice, place the burden of investigation on the developer as a condition, state specific investigation requirements or refuse permission depending on the available knowledge and assumed risk.

## INFORMATION REQUIRED FOR THE PLANNING PROCESS

Obviously there is a basic need for comprehensive and reliable information because, unless information is available there is a danger that decisions may be taken on development without the benefit of all relevant details. Armed with information better efforts can be made to cater adequately for development needs whilst being more aware of the risks.

By providing a means whereby preliminary decisions can be made on a choice or use of a site, with least anticipated hazard, it would be expected not only to save time and money but also to avoid expensive mistakes. It should also be possible to give some indication of the type and size of investigation required and the subsequent preventative remedial measures. At present the 'planner' does not have access to all the available information because geomorphological data are either not available on a systematic basis or because there is a general lack of awareness of the information that the geomorphologist can provide.

In most planning offices information on all forms of earth science constraints and potentials is assembled in an *ad hoc* way from available geological surveys; memoirs and papers; published geological maps of varying age, scale and accuracy; mine and quarry records; past site investigations or case records and local knowledge. It is always a desk study and therefore of variable, fragmentary quality. The level or reliability of the digest and interpretation of this information is only as good as the sources and the operators and must vary from Authority to Authority. In consequence geological constraints are often discovered too late and expensively in the development process. Land assessment is also carried out by Derelict Land Capital Programmes, land forums (e.g. West Midlands CC, 1982) or by joint land availability studies between local authorities and housebuilders' federations but these are not comprehensive (DoE, 1984).

In recognition of these problems the Department of the Environment through the British Geological Survey have commissioned, for certain key planning areas, Environmental Geology Maps (EGMs) whose aim is to portray those geological characteristics of the environment which should be taken into account during the planning process (DoE, 1984; Nickless, 1983; GSL, 1985).

These authoritative documents assemble and integrate existing geological and other selected data of relevance to each planning area (Table 7.1). They present the information in an easily accessible form and in non-technical language, but they include sufficient technical information to allow land decisions to be justified, a matter of importance during planning appeals.

**Table 7.1** Typical maps provided in a British environmental geology map

| | |
|---|---|
| Element maps: | Bedrock geology<br>Bedrock lithology<br>Superficial deposits<br>Thickness of superficial deposits<br>Rockhead contours<br>Sand and gravel deposits<br>Opencast workings<br>Shallow undermining<br>Limestone (or other economic material)<br>Landslides<br>Hydrogeology<br>Slope steepness<br>Location of borehole sites<br>(or site investigation areas) |
| Derived maps: | Underground storage potential<br>Sand and gravel potential<br>Foundation conditions<br>Groundwater resources |
| Potential maps: | Development potential<br>Priority areas for on-site investigations<br>Geological resources |
| Additional topics: | Extent and thickness of made ground<br>Engineering properties of bedrock<br>Engineering properties of superficial deposits<br>Geomorphology<br>Land use (and designated areas) |

A primary aim of the environmental geology surveys is to evaluate the significance of the information in terms of likely constraints on development and the resource potential of the area. They are therefore, broad in scale and attempt to integrate information across wide areas. They are not produced, however, to replace standard site investigations but to complement the normal information utilized by the planning process. At present there are fourteen studies for priority planning areas covering Glenrothes, Morley, Hurn-Christchurch, Aldridge-Brownhills, Rothwell, Glasgow, South Humberside, Edinburgh, Bridgend, Brierley Hill, London Dockland, Chacewater and West Wiltshire–SE Avon. A further report on Torbay is in progress.

With respect to landsliding there has been a determined effort by the Department of the Environment, the Scottish Development Authority and the Welsh Office to satisfy the urgent need for information on the distribution, nature and risk from landslides. This need is long established. For example, in the discussion of a famous paper by Knox (1927) on the landslides of South Wales a Mr L. Porcher, then Clerk to the Pontypridd Urban District Council said:

> One great lesson to be learnt from Professor Knox's paper was that no new townships should be established in any area until men of eminence in scientific circles had surveyed the area in order to foretell what was likely to happen and until the necessary remedial measures had been taken to avoid landslides. That was the only way to avoid great expenditure in the repair of avoidable damage in the future.

Unfortunately, this clear warning went unheeded and recent history has shown how prophetic it was. The Aberfan tragedy has, however, stimulated comprehensive official action in which the Department of the Environment, Welsh Office, the Institute of Geological Sciences and Sir William Halcrow and Partners have completed a pilot survey and catalogue of landslides in the South Wales coalfield area, detailed studies of selected landslides in Blainau–Gwent (e.g. the Bournville–East Pentwyn slides), and the development of a landslide hazard potential mapping scheme in the Rhondda area. The latter is the first comprehensive UK national hazard mapping methodology to be developed (Payne, 1985).

In Scotland a survey of rockfalls in the Highlands has been completed and the results of both this and the Welsh surveys have been incorporated in the state-of-the-art Review of Landsliding in Great Britain. As far as is known, however, there are no similar reviews for Northern Ireland. The landslide review is a DoE (Geological and Minerals Planning Research Programme) contract which seeks to establish what work has been done on landslides in the UK; to assess its applicability and to identify gaps in knowledge, priority areas and risk (DoE, 1986–7).

The project identifies the geographical distribution, causes and mechanisms of landslides, significant associations with causal factors, reviews methods of investigation, landslide potential mapping, remedial measures, legislative, and administrative provision, risk to communities and the need for advice to planners and future research. It revolutionizes our knowledge and awareness of the problem in Great Britain.

It might therefore seem that planning authorities are now well provided for. Yet when examined in detail these data sources do not fully answer the need. In planning offices *ad hoc* collations are perfectly acceptable if they are carried out by a specialist, if they are thorough and if the source materials are adequate. Rarely, however, are these needs satisfied. Most planning offices do not contain specialist geomorphologists capable of interpretation for the planning function. Existing data sources are often inadequate, unavoidable, out of date or the key may not be suitable for planning purposes.

Environmental geology maps are not available for all areas, are not of standardized format or content and, from a geomorphological–hydrological point of view are severely deficient. Usually missing are classifications of materials in terms of lithology or BS 5930 specifications, structural data or assessments of superficial deposits and their properties in terms relevant to the planner. Rarely do they contain morphological data such as slope steepness or details of specific landforms of diagnostic relevance (e.g. floodplains).

Present-day processes are almost entirely neglected and this includes surface hydrology, flood limits, landsliding activity, rockfall occurrence, coastal erosion, soil erosion and karst subsidence all of which are potential constraints to development. Environmental geology maps and reports are usually based only on existing data and their effectiveness is therefore restricted to the range of work already carried out in the area. It is obvious that this could be improved by the most elementary use of air photographs and fieldwork.

Neither should the Landslide Survey be seen as the complete answer to a planner's prayer. Certainly it is a massive step forward which, containing the details of over 7000 landslides, has become an essential work of reference for any planning, engineering or similar project. Nevertheless, it is only a review of published work showing the distributions of most *known* landslides on a scale of 1:250,000. It is intended to reveal what we do know as a guide to future research. There are, in fact, huge areas of unmapped terrain.

Geomorphologists are therefore faced with a great responsibility. The information required by the planner on which to base sensible decisions from an earth science–geomorphological viewpoint is simply not available. It is at least arguable that geomorphologists have a responsibility to provide these data. Certainly the current political climate expects science to provide *relevant* information and the DoE is providing a mechanism to achieve it.

In this context there is much to be learnt from overseas practice where a far more pragmatic view of research has been taken. In Europe, for example, there is a common attempt to produce:

—Geotechnical maps showing the suitability of land for different development functions and the conditions of investigation required (e.g. Abad, Del Moral and Pena Pinto, 1979; Aisenstein, Schulman and Israel 1974; Echevarria and Pena, 1978; Ghiste, 1980; Masson, 1971; Mazeus and du Mouza, 1979; Melidoro, 1970; Olivier and Camboly, 1979; Polo-Chiapolini, Schroeder and Monjoie, 1974; UNESCO, 1976).
—Resource maps (e.g. Bini and Del Sette, 1980; Mahr and Malgot, 1978).
—Risk maps, especially for seismic risk and landslides. Here the ZERMOS (*zones exposées aux risques liés aux mouvement du sol et du sous sol*) (e.g. Antoine, 1977; Champetier de Ribes, 1979; Chazan, 1973; Gandemer and Le Campian, 1979; Humbert, 1975) Figures 7.1(a),(b).
—The POS system—*Les plans d'occupation des sols* (e.g. Gounon, 1980).
—Environmental maps showing the joint opinions of earth scientists, planners and economists on land sensitive to impact (e.g. Burchell and Listokin, 1975; Dearman and Matula, 1976; Lee and Wood, 1978).

—Earth science data banks such as the FIDGI—*Fichiers de Données Géotechnique Informatique*—and VERCORS systems of France or the SYSFAP—*Système Intégré de Fichiers Auto Programmes*—of Liège (Buisson, 1976; Polo-Chiapolini, Schroeder and Monjoie, 1974).

The objectives, core topics and constraints to development included in surveys for planning in Europe are shown in Table 7.2 (a, b, c). Clearly geomorphology occupies a central part in such surveys.

In the USA there is also comprehensive legislative support to 'earth science for planning' via the National Environment Policy Act (1969) which allows for the provision of:

> Earth science information to planners in particular and to engineers, politicians and others concerned with development generally, in a useful form in order to improve design and development of land especially in the context of urban growth and development.

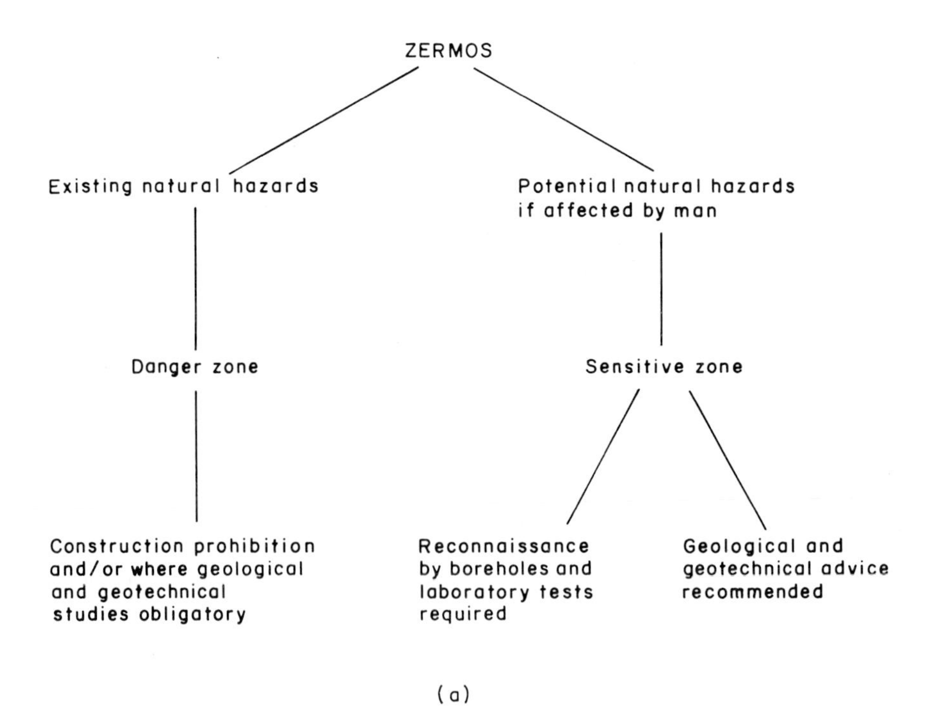


**Figure 7.1** (a) The French ZERMOS scheme for risk mapping (simplified). After Humbert, 1975. (b) Extract of a POS–ZERMOS map at a scale of 1/5000 of the Touques Valley, France

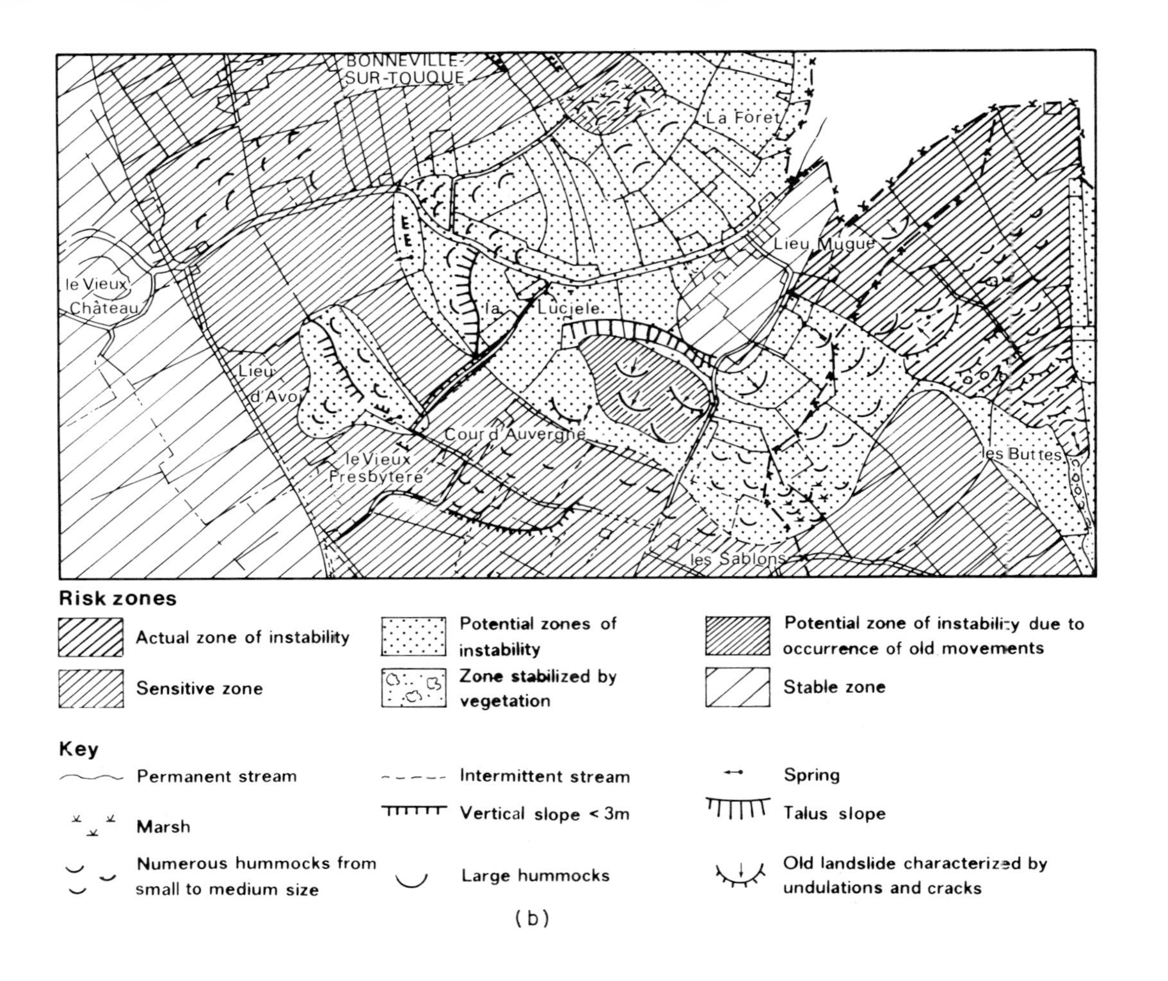
BONNEVILLE-SUR-TOUQUE
La Foret
Lieu Mugue
le Vieux Château
la Luciele
Lieu d'Avo
Cour d'Auvergne
le Vieux Presbytere
les Buttes
les Sablons
Risk zones
Actual zone of instability
Sensitive zone
Potential zones of instability
Zone stabilized by vegetation
Potential zone of instability due to occurrence of old movements
Stable zone
Key
Permanent stream
Marsh
Numerous hummocks from small to medium size
Intermittent stream
Vertical slope < 3m
Large hummocks
Spring
Talus slope
Old landslide characterized by undulations and cracks

(b)

**Table 7.2 (a)** The specific objectives of Environmental Geology Maps as practised in Europe

| | |
|---|---|
| To categorize | —stratigraphy, lithology, structure bedrock and superficial deposits in geotechnical terms<br>—the geo-resources<br>—the natural hazards<br>—precise quantitative measures of the ground and surface process systems.<br>—the spatial and temporal variability of the natural processes |
| To evaluate | —competing land uses suitable for each terrain area<br>—potential environmental damage<br>—the advantages and disadvantages of each site so that the best development may be selected<br>—the most economic solution |
| To inform | —by disseminating sound information in a form suitable for use by the planner, engineer and public<br>—by calling attention to resources, risks poor conditions<br>—by creating data-banks suitable for reference, prediction, updating<br>—offering specific advice on special problems |
| To promote | —sensible methods of survey<br>—sound practice<br>—legislation, planning law, ordinances, codes of practice, structure plans, etc. |

**Table 7.2 (b)** Core topics on European Environmental Geology Maps

| Topics | Common subdivisions |
|---|---|
| Geomorphology | Rockfall, debris chutes, debris avalanches, landslides, creep, solifluction, subsidence, erosion, flooding, karst, drainage, permeability. Usually classified as geodynamic risks (scale, intensity, rate, probability, age). |
| Morphology | Slope gradient, slope classification, drainage parameters, terraces, ridges. Heights. |
| Stratigraphy | Usually standard geological column division. Bedrock and superficial. European maps emphasize a wide superficial division (e.g. terraces, alluvium, colluvium, scree, glacis, solifluction debris). Always includes thickness and depth to formation. |
| Lithology | Usually a detailed division of particle size categories, occasionally roundness, durability, chemistry. |
| Structure | Dip, strike, thickness, faults, active, inactive, thrust, bedding planes, fracture spacing, jointing. |
| Tectonic/Seismic | Activity of faults, microseismic zones, epicentres, intensity of state of rock, fracture indices, bedding form related to tectonics. |
| Hydrogeology | Depth to aquifer, number and type. Hydrostatic condition, piezometry, permeability of rock or soil, potability, chemistry, pollution susceptibility. |

*(continued)*

**Table 7.2 (b)** *continued*

| Topics | Common subdivisions |
|---|---|
| Surface hydrology | Water resources, run-off, discharge, sources, sinks, drainage pattern density, infiltration capacity, climatic/hydrograph summaries. |
| Geotechnical data | Compression, point load, bearing capacity, penetration, strength. Descriptive properties, Atterberg limits, pore-pressure details. |
| Location data | Grid coordinates. |
| Special conditions | For example, Karst, caves, underground workings, mines, cavities, soluble materials. Weathering state, grade, process. |
| Anthropogenetic features | Quarries, workings. |
| Protective works | Check dams, lined channels, dams, structures, gauging stations, retaining walls. Stabilization measures. |
| Vegetation | Type, area, cover. |
| Land use | Urban, rural, arable, crops, vines, etc. |
| Pedological soils | Subdivision type, use or workability. Grade. |
| Ecological landscape | Sites of scientific interest, unique sites, resources, protected areas, parks, etc. |
| Archaeological | Sites of archaeological interest, monuments, castles, graves, etc. |
| Location or original data | Maps, data, archives, aerial photographs. Historical. |

**Table 7.2 (c)** Geomorphological constraints in development shown on European Environmental Geology Maps

| |
|---|
| Steep slopes |
| Land liable to landsliding |
| Ground instability caused by subsidence, solution and piping |
| Erosion by sea, rivers, channel change |
| Flooding or waterlogging |
| Desiccation or cracking |
| Hydrogeology—permeability |
| Siltation |
| Aeolian processes—dune movement |
| Saline ground |
| Distribution, thickness etc.—made ground |
| Location of household, toxic, hazardous waste or fill |
| Metastable soils |
| Areas sensitive to change, channel change, headward erosion, slope failure, coast recession (direction and rates) |
| Areas sensitive if disturbed |

In the USA 'earth science' is broadly-interpreted and EGMs are almost as dependent on geomorphological/hydrological data as on geology. They are particularly advanced in the production of hazard maps (e.g. Christenson *et al.*, 1978/9; Glaser, 1976; Jacobs, 1971; Robinson and Spicker, 1978; Cooke, 1984; Varnes, 1984).

## CONCLUSION

In addition to learning from overseas experience it is necessary in the UK to encourage Government to develop a General Policy Statement, Legislation and Method of Approach to the assessment of landslips and to other earth science constraints to development. These should be written for the guidance of planners

**Table 7.3** Methodology for the preparation of Landscape Sensitivity Maps for planning purposes (Brunsden, 1987)

| *Step* | | |
|---|---|---|
| 1 | Prepare | Factual base maps. Topography. Geology. Structure geomorphology. Hydrology, Soils. |
| 2 | Tabulate | Large scale cycles, epicycles, episodes of landscape evolution. |
| 3 | Collate | Evidence on dates, frequencies, magnitudes, durations of climatic, seismic, flooding, landslide, etc. events. |
| 4 | Determine | Probabilities and thresholds for each process. |
| 5 | Classify | Vulnerable classes of landscape (e.g. 1:50 event causes failure of >35° slope on slate). |
| 6 | Devise | A scheme of proxy variables, using statistical analysis to enable the probability values or recurrence intervals to be assigned to the classified landscape areas. |
| 7 | Determine | The diagnostic large events which cause landscape change using the indicators of past events as a guide to future risk. |
| 8 | Maps | The nature redirection of aggressive erosional stress (e.g. erosion propagation). |
| 9 | Maps | The variable spatial sensitivity or resistance to change (e.g. flat slopes, distance from erosion axis, low drainage density, resistant rock). |
| 10 | Isolate | Vulnerable rock types, swelling clay, shattered rocks. Metastable soils, etc. |
| 11 | Predict | Worst state (e.g. saturated ground). |
| 12 | Maps | Any known threshold relationships (e.g. slope angle *v* slope failure, transitional sliding *v* regolith thickness *v* shape). |
| 13 | Utilize | Post-audit surveys and learn from mistakes by examining what happened in particular events or situations in the past. |

in the determination of land use allocation in local plans, in the handling of planning applications and in the timing of the application of expert advice.

To assist this we must develop a Code of Practice similar to a British Standard or a Geological Society Recommended Practice Report, on the correct procedures for the application of geomorphology to planning. The code of practice should contain, or develop, a methodology for factual maps, data analyses for the recognition of palaeo-conditions (e.g. late glacial slip surfaces), form (e.g. slope steepness), process (e.g. landslides, coast erosion, siltation), hydrology (e.g. channel change, flood limits), material properties (e.g. density, strength), classifications of superficial deposits and their properties and any special problems such as karst, metastable soils, swelling soils and subsidence.

The final reports should include statements on the constraints to development (Figure 7.2c) and resources. There is also a real demand to map comprehensively the country and to develop suitable computerized data banks or data source banks according to an agreed standard code of practice on landslide data collection techniques, hazard mapping methodology, prediction and assessment of risk.

In the future and more exciting for the research field it is necessary to apply the latest theoretical work to the problem. The recent development of landscape sensitivity maps (Brunsden and Thornes, 1979; Brunsden, 1987) for example, is based on the actual characteristics of landform evolution in a particular area and founded on relevant principles of landform change rather than on abstract general evolutionary models or artificial and generally unsubstantiated hazard mapping methodologies (Table 7.3).

A further approach is to develop the idea of landform design in order to create 'natural' landforms capable of long-term stability. This is necessary in order to minimize the cost of long-term maintenance, to ensure long-term hazardous waste disposal, to create 'slow-rate' processes and to design minimum erosion slope systems. Similarly we need to use our knowledge of recurrence intervals of events and relaxation times in order to short-circuit natural systems and create equilibrium forms 'before their time'.

The challenge of imaginative, creative, geomorphological design and engineering is before us.

## REFERENCES

Abad, J., J. Del Moral and J. L. Pena Pinto (1979). Spanish experience of geotechnical cartography in an urban area, *Bull. Int. Assoc. Eng. Geol.*, **19**, 79–84.

Aisenstein, B., N. Schulman and A. Israel (1974). The geotechnical maps of Jerusalem, *II. Congress I.A.E.G., Sao Paulo*, 1, T9.

Antoine, P. (1977). Reflexions sur la cartographie ZERMOS et bilan des experiences en cours, *Bull. B.R.G.M.*, 2 e Levie, Sect III, n 1/2, 9–20.

Bini, C. and Del Sette, M. (1980). The contribution of soil cartography to land use planning, *Bull. Int. Assoc. Eng. Geol.*, **22**. 151–6.

British Standards Institution (1981). *Code of Practice for Site Investigations*, BS 5930, London, 174pp.
Brunsden, D. (1987). Principles of hazard assessment in neotectonic terrains, *Proc. 1st. Sino-British Geological Conference, 1-9 April 1987*, Taiwan.
Brunsden, D., J. C. Doornkamp and D. K. C. Jones (1978). Applied geomorphology: a British view. In Embleton, C., D. Brunsden and D. K. C. Jones (eds), *Geomorphology, Present Problems and Future Prospects*, OUP, pp. 251–62.
Brunsden, D., J. C. Doornkamp and D. K. C. Jones (1979). The Bahrain Surface Materials Resources Survey and its application to regional planning, *Geographical Journal*, **145**, 1-35.
Brunsden, D. and J. B. Thornes (1979). Landscape sensitivity and change, *Trans. Inst. Brit. Geogrs.*, New Series **4**(4), 463–84.
Buisson, J. L. (1976). Le fichier des données géotechniques, *Bull. de Liason des Laboratoire des Ponts et Chaussées*, **84**, 149–62.
Burchell, R. W. and D. Listokin (1975). *The Environmental Impact Handbook*, Rutgers Univ., NJ, Centre for Urban Policy Research, 200pp.
Champetier de Ribes, G. (1979). Données géologiques et géotechniques et plans d'occupation des sols, *Bull. de Liason des Laboratoires des Ponts et Chaussées*, 2277–8, 99.
Chazan, W. (1973). Le plan ZERMOS; identification des zones exposées aux mouvements du sol et du sous-sol, prealable à leur prevision et á la prevention de leur effets, *Symp. Nat. S.S.S.C.*, Cannes, 93-9.
Christenson, G. E., T. L. Péwé and others (1978/9). Environmental Geology of the McDowell Mountains Area, Maricopa County, Arizona, *State of Arizona, Bur. of Geol. and Min. Technology*, Geol. Investigation Series, G.1.
Clarke, M. L. (1985). Mid-Glamorgan—a case study. Rhondda—implications for a District Authority, In Morgan, C. S. (ed.) *Landslides in the South Wales Coalfield*, Proc. Symp. Poly. Wales, Civ. Eng. and Building, 1-3 April, 153–69.
Cooke, R. U. (1984). *Geomorphological Hazards in Los Angeles*, Allen & Unwin, 187pp.
Cooke, R. U., D. Brunsden, J. C. Doornkamp and D. K. C. Jones (1982). *Urban Geomorphology in Drylands*, Oxford U. Press, 324pp.
Dearman, W. R. and M. Matula (1976). Environmental aspects of engineering geological mapping, *Bull. Int. Assoc. Eng. Geol.*, **14**, 141–6.
Department of the Environment (1984). *Land for Housing*, circ. 15/84, HMSO, London.
Department of the Environment (1986–7). *Review of Landsliding in Great Britain* (Unpublished).
Doornkamp, J. C., D. Brunsden, D. K. C. Jones, R. U. Cooke and P. R. Bush (1979). Rapid geomorphological assessments for engineering, *Quart. Journ. Eng. Geol.*, **12**, 189–204.
Echevarria, C. M. R. and J. L. Pena (1978). La cartografia geotécnia en Espana, *Tecniterrae*, **25**, 19–27.
Flynn, M. R. (1985). Planning and development implications, In Morgan, C. S. (ed.), *Landslides in the South Wales Coalfield*, Proc. Symp., Poly. Wales, Civ. Eng. and Building, 1-3 April 1985, 125–130.
Gandemer, P. H. and M. Le Campian (1979). Cartographie des risques ZERMOS appliquées a des plans d'occupation des sols en Normandie, *Bull de Liason des Laboratoires des Ponts et Chaussées*, **99**, 2277, 43–4.
Geomorphological Services Ltd (1985). *Environmental Geology Mapping*, Report to DoE, 2 vols, 73pp. and 136pp.
Ghiste, S. (1980). Facteurs anthropologiques et cartographie géotechnique, *Bull. Int. Assoc. Eng. Geol.*, **22**, 125–8.

Glaser, J. D. (1976). *Anne Arundel County: Geology, Mineral Resources, Land Modifications and Shoreline Conditions*, Maryland Geological Survey, County Atlas No. 1.

Gounon, A. (1980). La planification urbaine basée sur la géologie l'hydrogéologie, la géotechnique. Un example: la Ville de Nice, *Bull. Int. Assoc. Eng. Geol.*, **22**, 161–6.

Humbert, M. (1975). Establissement des cartes de localisation probable des zones exposées à des mouvements du sol cartographie, *ZERMOS 75 SGN 127, AME, BRGM*, Orleans.

Jacobs, A. M. (Comp.) (1971). *Geology for Planning in St. Clair County, Illinois*, Illinois State Geological Survey, Circ. 165.

Jones, D. K. C. (1980). British applied geomorphology: an appraisal, *Zeits. für Geom.*, Supp. 36, 48–73.

Knox, G. (1927). Landslides in the South Wales Valleys, *Proc. Trans. South Wales Inst. Engrs.*, **43**, 161–247 and 257–90.

Lambert, D. G. (1985). Aspects of the Legal Situation—liability and responsibility, In C. S. Morgan (ed). *Landslides in the South Wales Coalfield*, Proc. Symp. Poly. Wales, Civ. Eng. and Building, 1–3 April 1985, 131–8.

Lee, N. and C. Wood (1978). Environmental impact assessment of projects in E.E.C. countries, *Journ. Environ. Management*, **6**, 57–71.

Mahr, T. and J. Malgot (1978). Zoning maps for regional and urban development based on slope stability, *Int. Assoc. Eng. Geol.*, 3rd Int. Eng., **1**(1), 124–37.

Masson, M. (1971). Cartographie géotechnique de l'Agglomeration Rouenaise, *Bull. de Liason de Laboratoire des Ponts et Chaussées*, **52**, 109–17.

Mazeus, H. and du Mouza, J. (1979). Cartographie géotechnique de formations superficielles en zones non urbanisées, *Bull. Int. Assoc. Eng. Geol.*, **19**, 47–52.

Melidoro, G. (1970). La Cartografia Geotécnica principi, experience e reflession, *Geol. Appl. Idrog.*, **5**, 27–47.

Nickless, E. (1983). Environmental geology mapping, *Mineral Planning*, **17**, 25–9.

Olivier, G. and D. Camboly (1979). Cartographie géotechnique pour l'aménagement, *Bull de Liason de Laboratoire des Ponts et Chaussées*, 2278, 56–64.

Payne, H. R. (1985). Hazard assessment and rating methods. In Morgan, C. S. (ed.) *Landslides in the South Wales Coalfield*, Proc. Symp. Poly. Wales, Civ. Eng. and Building, 1–3 April 1985, 131–8.

Polo-Chiapolini, C., C. Schroeder and A. Monjoie (1974). Cartographie Géotechnique Automatique du Centre de Liège et du Sart Tilman, *II, Int. Lang. ITEG*, Sao Paulo.

Robinson, G. D. and A. M. Spicker (eds) (1978). *Nature to be Commanded: Earth Science Maps Applied to Land and Water Management*, US Geological Survey, Professional Paper, 950.

UNESCO (1976). Guide pour la préparation des cartes géotechnique, *Inst. Assoc. Eng. Geol., Mapping Commission*, Paris, 78pp.

Varnes, D. J. (1984). *Landslide hazard zonation: a review of principles and practice*, UNESCO, 1984, 63pp.

Geomorphology in Environmental Planning
Edited by J. M. Hooke



# 8 A Geomorphological Approach to Limestone Quarry Restoration

**PETER GAGEN and JOHN GUNN**
*Limestone Research Group, Department of Environmental and Geographical Studies, Manchester Polytechnic*

## INTRODUCTION

Limestone is an essential raw material for any industrial nation. It is used in the manufacture of iron and steel, cement, glass, chemicals, ceramics, fertilizers, plastics, paints, paper; the refining of basic footstuffs including sugar and flour; as an aggregate in the construction of roads and buildings; in agriculture for soil treatment and as a nutritional input to animal foodstuffs; and in the purification of water supplies and effluent management. Limestone and lime are required at some stage, either directly or indirectly, as primary or allied ingredients, in a diverse range of manufactured products. The limestone quarrying industry is concerned with the production of sufficient quantity of this raw material to meet the demands associated with these multifarious needs.

Limestone quarrying in the United Kingdom is concentrated on the Carboniferous Limestone which outcrops mainly in the Peak District, the Mendip Hills, Gloucestershire, North Yorkshire, Cumbria, North and South Wales and Northern Ireland. Also of importance are the Permian limestone and the Cretaceous limestone or chalk which is used extensively for cement making. National limestone production in 1979 was 89.2 million tonnes (Harrison, 1981). Approximately 25 per cent of production is from the Peak District of Derbyshire and Staffordshire followed by 14 per cent from the Mendip Hills.

The exploitation of the most accessible deposits in large open quarries of up to 5 million tonnes annual capacity can conflict with the attractiveness of the countryside associated with these limestone outcrops as, for example, in the Peak District and Yorkshire Dales National Parks. The reconciliation of competing demands on the landscape, particularly the national need for limestone and conservation of the countryside, requires the formulation of policies encompassing initial exploration, development control, restoration and after-use of quarry workings.

The research described in this chapter commenced as a study of the geomorphological implications of limestone quarrying as a human agency of landform change. One of the products was a model of landform evolution on quarried limestone rock slopes which is described below. From discussions with mineral operators and mineral planners it became apparent that present methods for limestone quarry rehabilitation are inadequate and that the research being undertaken could be applied to this problem. As a result a theory for the construction of skeletal rock landforms by 'restoration blasting' was developed. This technique of drilling and blasting is designed to restore quarried rock slopes to a sequence of landforms which not only mimic the outward form of those of a natural limestone daleside but which can be predicted to evolve in harmony with the continued operation of natural processes. The potential for landform reconstruction by restoration blasting has in turn led to a reconsideration of existing legislation governing the rehabilitation and after-use of limestone quarry workings.

## LIMESTONE QUARRYING AS A GEOMORPHIC PROCESS

The majority of geomorphological studies of 'man's impact on the environment' have focused on ways in which human activities impinge upon natural processes (thereby altering their rate of operation) and natural landforms (thereby altering their form). In contrast surprisingly little has been written on the rate and impact of direct human erosion where materials are broken down, often by the use of explosives, and removed by machinery. The first, and still the most comprehensive, attempt to quantify 'human denudation' in the United Kingdom was made by Sherlock (1922) in his seminal book on *Man as a Geological Agent*. It is of interest to note that limestone quarrying did not receive separate consideration from Sherlock, probably on account of the relatively minor scale of the industry in the early part of the present century. Hence, it was left to Dearden (1963) to make the first attempt to quantify the removal of limestone from the Peak District 'by man and nature'. He concluded, on the basis of data for 1954, that human actions were about seventy times more rapid than natural. Since 1954 the rate of limestone extraction has increased by more than 300 per cent reaching a peak during the period 1969–1983 (Figure 8.1). In that fifteen-year period over 286 million tonnes of limestone were removed from the Peak District by quarrying while less than 15 million tonnes were removed in solution (Gunn, Gagen and Raper, in preparation). It has also been estimated that by the end of the present century a similar volume of limestone will have been removed from the Peak District by direct human erosion (quarrying) as has left in solution during the Holocene (Gunn, Gagen and Raper, in preparation).

However, quarrying is spatially concentrated so that the extent of the increased erosion and its geomorphological impact varies between drainage basins.

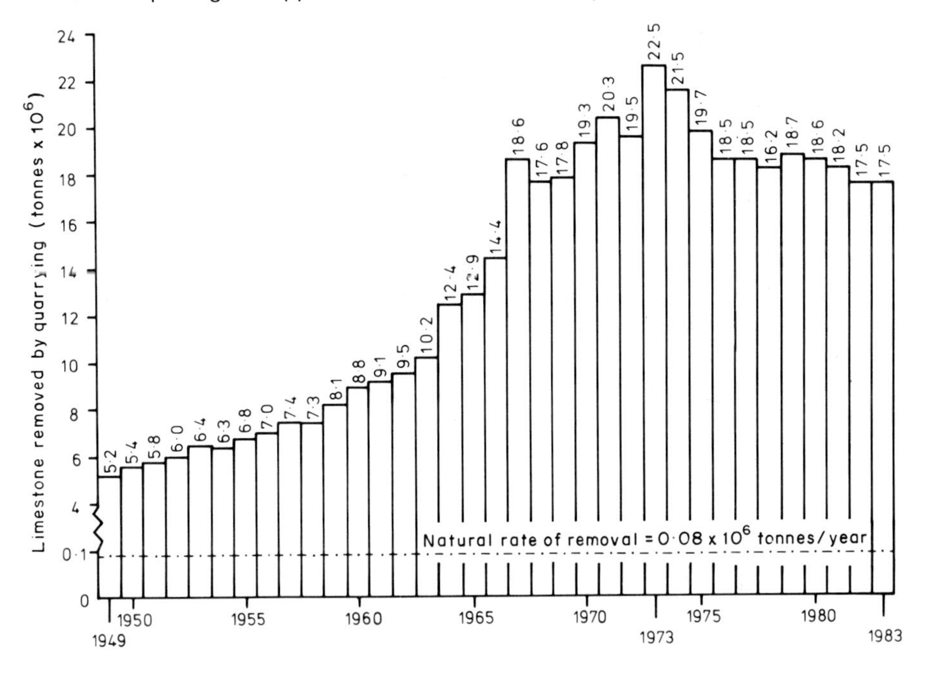


**Figure 8.1** Limestone production in the Peak District, 1949–83

The two most obvious impacts are the destruction of certain landforms, notably caves and closed depressions (dolines/sinkholes) and the substantial modification of others, notably dalesides and increasingly entire hills. These impacts may be sufficiently severe as to warrant refusal of planning permission for further extraction as at Eldon Hill Quarry near Castleton, Derbyshire. However, in many instances the main objections to quarrying result from the conspicuous, engineered appearance of the quarried rock faces which remain after working has ceased. From a geomorphological perspective these faces, and the quarries of which they are a part, may be viewed as created landforms which will subsequently evolve under the influence of natural processes. They are an appropriate subject for geomorphological research although little work of this kind has been previously undertaken, the only comparable study being that of Haigh (1978) who examined the evolution of slopes on constructional landforms which resulted from the mining of coal in South Wales.

## STUDY AREA

The Peak District is situated at the southern end of the Pennine range of hills, a broad anticline of Carboniferous rocks with its crest eroded such that the oldest strata now outcrop at its core. Limestones outcrop over an area of 450 km$^2$ known as the White Peak. This is essentially a soil covered, gently undulating plateau ranging in altitude from 275–450 m which is pitted by sinkholes and dissected by a complex network of largely dry valleys (Gunn, 1985). The technical ability of human beings to excavate, process and transport large tonnages of rock aggregate is well-represented within the White Peak. The quarrying of limestone has a long history in the area and is now the dominant extractive industry (Gunn, Hardman and Lindesey, 1985). It has grown from localized extractions using hand tools and the horse and cart, to the present day use of explosives, pneumatic drilling rigs, mechanical excavators and road and rail transport.

Eleven abandoned limestone quarries together with recently worked faces in Tunstead Quarry were investigated in Great Rocks Dale. This dry valley lies 3 km

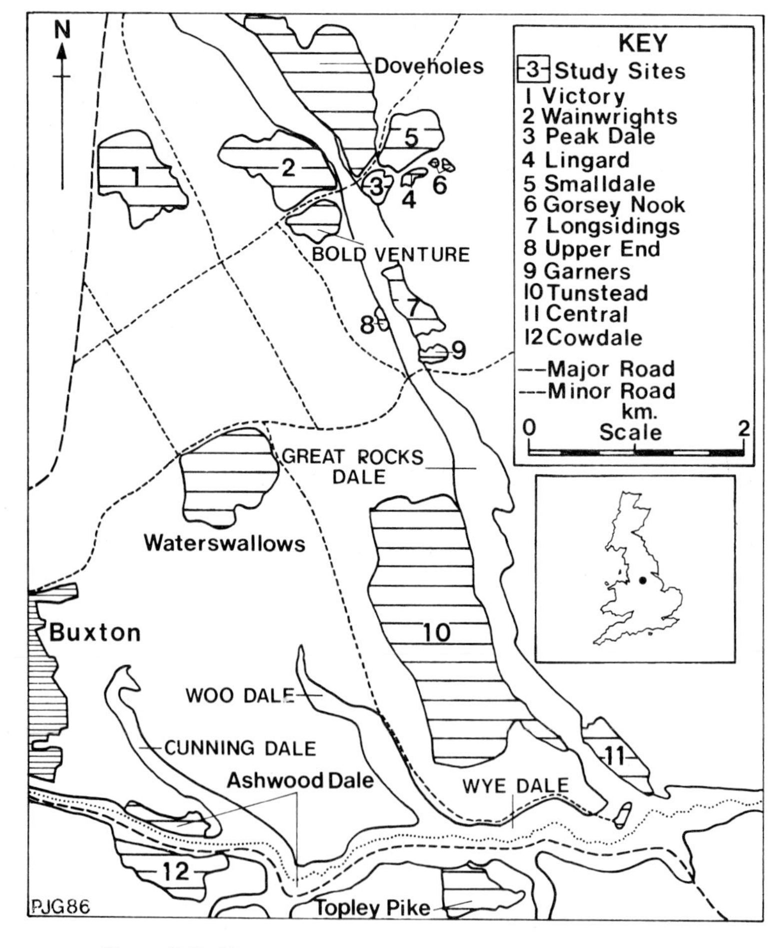


Figure 8.2 Limestone quarry study sites, Great Rocks Dale

east of Buxton and runs for some 5 km north-west to south-east from Doveholes at its northernmost end to its junction with the River Wye in the south (Figure 8.2). A total of 22 quarries worked the dale in the early 1920s but only two are now active, Doveholes in the north and Tunstead 3 km south. Many of the earlier quarries were abandoned following the take-over or amalgamation of smaller operators as the most expedient way of ending active competition. They then became part of the reserves of the Buxton Lime Firms with its establishment in 1928 and this subsequently gave rise to the formation of the Lime Division of ICI plc, the current operator of the largest working quarry in the Peak District at Tunstead.

### Site Selection

The criteria for quarry selection were (1) age—time since active working ceased; (2) methods of working—particularly drilling and blasting design; (3) size—area of working and height of worked faces; and (4) overall end-form of the worked area and its topographic situation within the dale. These requirements were met, together with the provision of all records relating to the working life of each of the quarries, by ICI plc who currently own and previously worked all of the sites investigated. It was therefore possible to establish the temporal sequence of rock-faces excavated within each of the sites, together with the dates at which extraction commenced and ceased. Two kilometres of limestone quarry rock-face were studied incorporating the principal geological divisions of the limestone outcrop within the study area and ranging in time since abandonment from two to over eighty years.

## LANDFORMS OF LIMESTONE QUARRYING

In common with most of the limestone outcrops in Britain, the earliest limestone quarries in the Peak District were small, shallow holes in the ground or locally exposed rock cuttings. In contrast modern workings consist of extensive and multiple extraction faces with exposures of rock of up to a kilometre in extent and individual rock faces of 20 m or more in height. Blasting operations can be carried out at a series of benched levels which may ultimately descend for over 100 m from the original ground surface. Extraction operations are largely continuous and the removal of stone is only rarely limited by technically insurmountable conditions. The ability of quarrying operations to create landforms is evidenced by the excavation itself which produces increasingly large quarry rock basins. The largest of these in the Peak District is Tunstead Quarry which extends over approximately 4 $km^2$.

Under present environmental conditions natural rock slopes are often regarded as essentially stable landforms which have reached a characteristic equilibrium form. Quarried rock slopes are often perceived to be similarly stable and unlikely to alter greatly from their excavated form by virtue of their engineered origin. However, geomorphological principles suggest that these rock slopes should be regarded as young landforms which are out of equilibrium with their surrounding environment and therefore likely to evolve rapidly from their form on abandonment (Thornes and Brunsden, 1977). Hence, it was decided to monitor

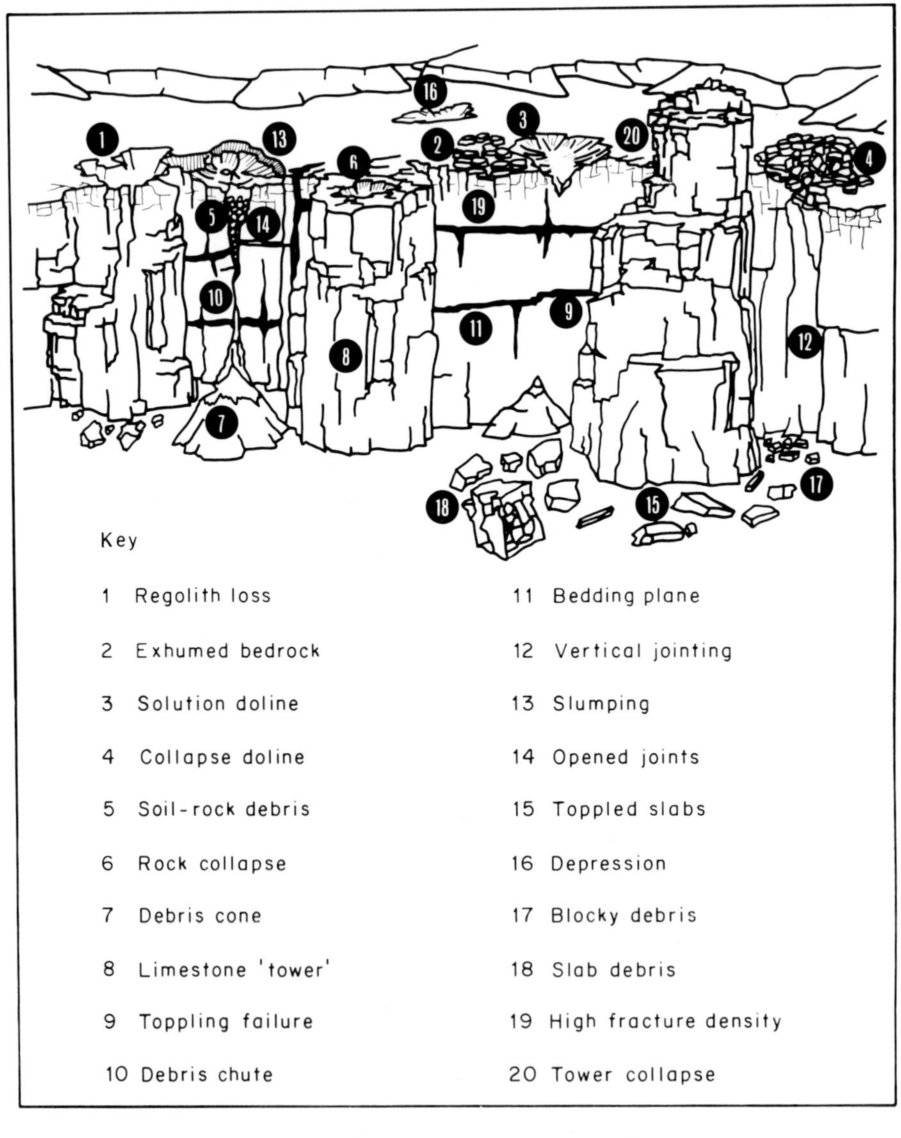


**Figure 8.3** A model of limestone quarry landforms

rockfalls from quarry faces with a range of ages since abandonment. A total of 120 1 $m^2$ debris-collection traps were installed beneath faces in the twelve quarries. Rockfalls were collected for twelve months at fortnightly, monthly and three-monthly intervals for individual and groups of quarries with the principal axis, weight and shape category recorded for the trap contents. The nature, magnitude and frequency of rockfall evidenced substantial changes in the form of the quarry faces over time. Rockfall was found to vary in response to the selective action of solutional and mechanical processes of weathering and erosion operating over the quarry faces. Rock slope recession was found to occur at different rates across faces of the same age and to vary in magnitude with the age of the quarry face.

Observations of these changes and of the general form of quarried rock slopes led to the identification of a suite of landforms analogous to those of natural limestone dalesides (Gagen and Gunn, 1987). These include limestone towers (rock buttresses), sinkholes (collapse dolines), sinkhole-like features (blast fracture cones), and rock debris chutes, cones and flows (Figure 8.3). These landforms can be divided into those which result directly from quarrying operations and those which are modified by quarrying operations.

## Landforms Resulting Directly From Quarrying Operations

The three principal landforms resulting directly from quarrying operations are blast fracture cones, rock buttresses and rock debris chutes, cones or flows. The term blast fracture cone is applied to both cone-shaped areas of fracturing which taper out down the rock face and to the roughly semicircular features which are a result of collapse from these same areas of fracturing (Figure 8.4a). The cones are sinkhole-like features with a lateral extent of 3–5 m which occur at regular intervals along the upper third of the total face height. Rock buttresses project out from the quarry face and increase in size and lateral extent towards the quarry or bench floor (Figure 8.4b). They develop alternately between blast fracture cones and are largest on older worked faces where they occupy approximately two-thirds of the total face height. They taper up the rock-face, disappearing as definite features at the same level as the apex of blast fracture cones. They were observed in various stages of collapse where rock sliding, toppling and slab failures occurred due to loss of support at each side of their upper portions following collapse of blast fracture cones. The collapse and generation of rockfalls from blast fracture cones is often augmented by wide vertical joints which channel rock, soil and clay material down the quarry face. These debris chutes produce a series of cones of rockfall material at the foot of the quarry face. Where this material is augmented by rockfalls from the rock buttresses substantial debris flows can be mobilized during wet weather. This was particularly apparent where groundwater issued from the rock-face across widened bedding plane surfaces and open joints.

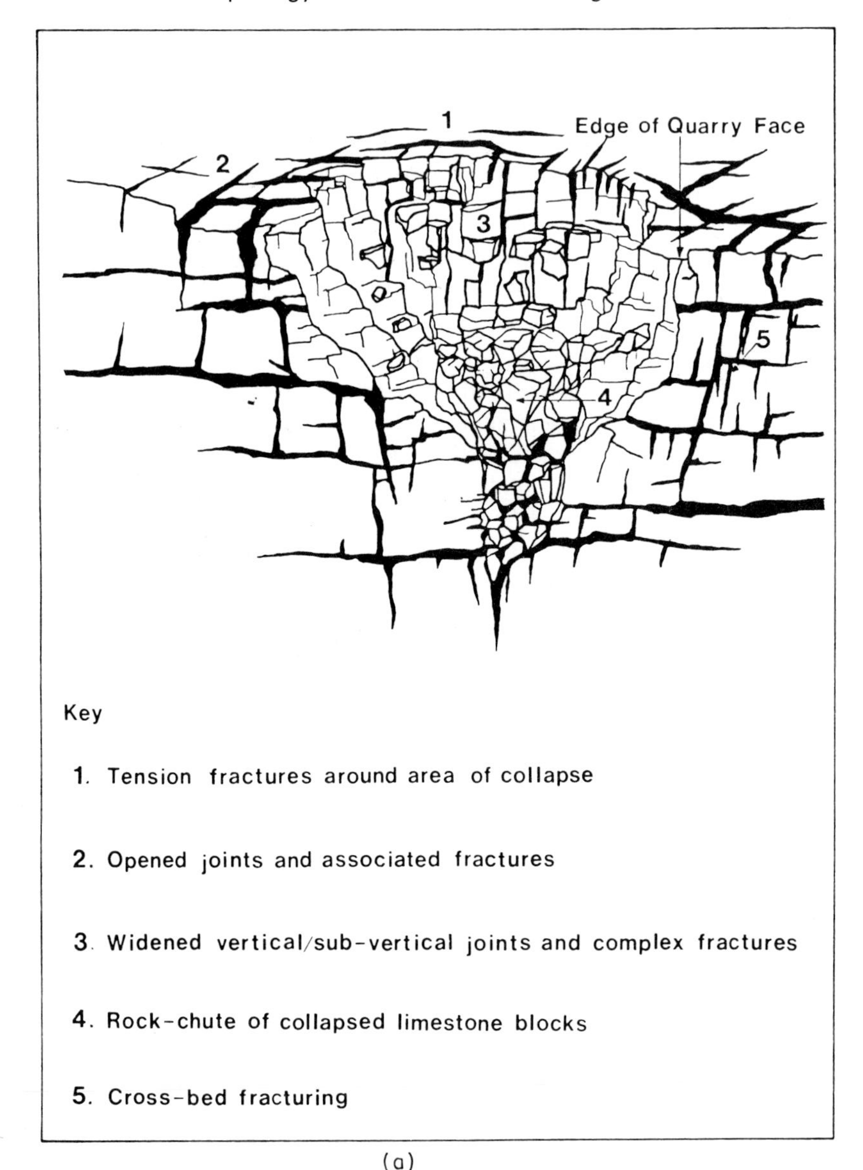


(a)

**Figure 8.4** Diagram of (a) blast fracture cone and (b) rock buttress

## Landforms Modified by Quarrying Operations

Sinkholes occur in areas of cleared ground around the margins of abandoned quarries. Such areas of 'piked' ground were cleared by gangs of quarrymen using iron piking rods. This removed the soil and clay overburden prior to blasting and so reduced contamination of the blasted stone. It also revealed the presence

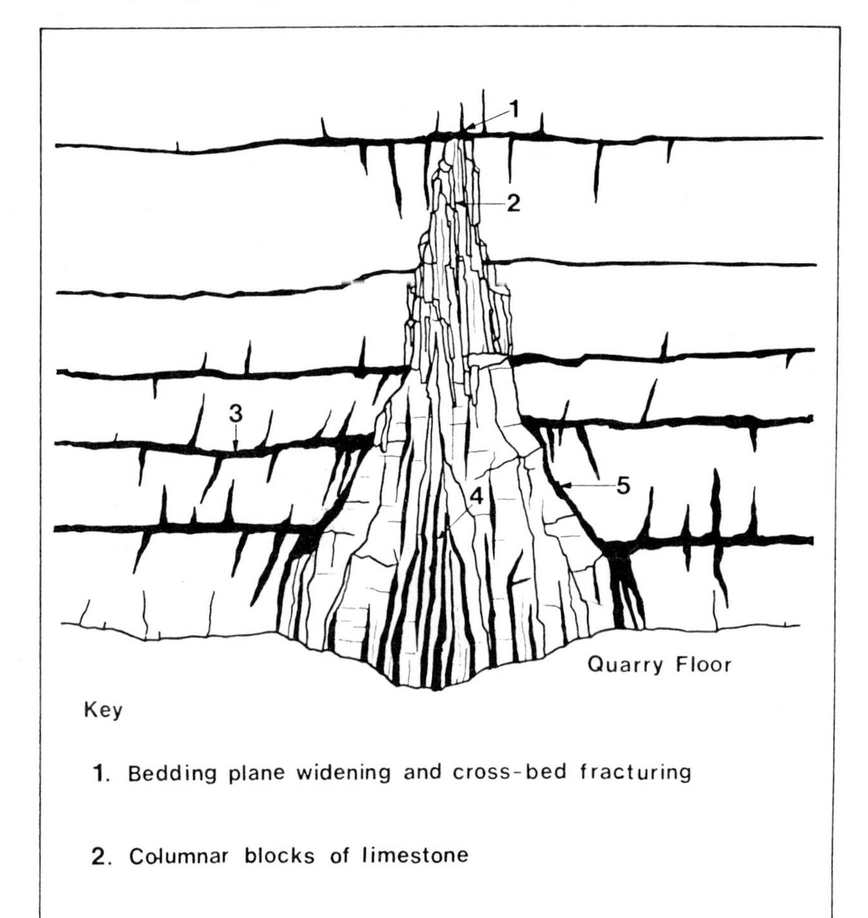


(b)

of joints and natural fractures, the deepest of which were selected for charging with blackpowder explosive to augment charges placed in header tunnels at the foot of the quarry face. The cleared ground would extend back from the face to be blasted for up to 10 m and run parallel with it for the length of face to be worked. Many of the quarries excavated prior to the 1930s possess this cleared ground and it is within these areas that sinkholes in various stages of development are found. The size and degree of development of these sinkholes decreases back from the edge of the quarry face, the widest and deepest features occurring within a metre of the edge. Evidence of natural sinkholes which have been intersected by excavation of the quarry face and which have collapsed as a result is seen in the presence of semi-circular cuts back into the previously straight line of the face. Areas of subsidence of up to 5 m across and 3 m deep were found within the piked ground (Gunn and Gagen, 1987). Today the overburden is cleared prior to drilling and blasting by mechanical excavators which also remove the underlying, weathered limestone leaving an almost bare rock surface.

## CONTROLS ON QUARRY LANDFORM DEVELOPMENT

The problems which inhibit the use of geomorphological evidence to determine the evolution of natural rock slopes were summarized by Thornes and Brunsden (1977, p. 23) as

> an inability to determine initial and boundary conditions; that when initial conditions are verifiable they almost invariably have to be plane surfaces because any less regular surface can never be determined with sufficient accuracy; that our knowledge of past conditions falls far short of the accuracy needed for significant assessment

and 'that any theoretical problem suitable for a comparison with those of the real landscape will be sufficiently complex to rule out any solution by analytical methods'. These difficulties are considerably reduced if quarry rock slopes are studied because they possess attributes which make them singularly suitable as anthropogenic landforms for an investigation into rock slope development over time: they can be accurately dated; the angle to which the slope was last excavated is a part of quarry company records; their mode of origin is well-documented and the methods of excavation are recorded. Furthermore, the present boundaries of quarried slopes can be readily and accurately determined, whilst previous topographic and site information was available from the quarry company. In order to determine the evolution of the quarry rock slopes it was first necessary to examine the methods by which they had been created. This involved consideration of blasting design (which includes explosives, drilling techniques and charging of shot-holes) and fracture patterns.

### Blasting Design

The twelve quarries investigated contain faces whose date of last working ranges from the late 1890s to 1984. During this period there have been two principal and several allied changes in the methods of quarry excavation as a result of technical progress in rock drilling and explosives blasting. Whilst many of these changes have been effected gradually one in particular is significant for its impact on both rock extraction and the generation of landforms over excavated rock-faces, and that is blasting practice. The blasting or 'getting' of stone has radically altered over the past fifty years and differs greatly from the first days of quarrying in the White Peak, especially with regard to the nature of the explosives used.

The choice of explosives used regulates the type of blasting design which can be adopted. Three main types of explosives are used: low explosive (e.g. blackpowder), high explosive (e.g. nitroglycerine and trinitrotoluene TNT) and ANFO a mixture of ammonium nitrate and fuel-oil. Blackpowder was used extensively throughout the early quarrying industry prior to the widespread introduction in 1949 of high explosives. Since the late 1960s high explosives have increasingly been used in conjunction with ANFO. The essential property of any explosive is that, on detonation, it is converted as rapidly as possible into gases which occupy many times the original volume of the explosive. In high explosives the gases are produced almost instantaneously at very high temperatures and pressures and are accompanied by an intense shock-wave. Blackpowder is slower in action and the gases are released at much lower pressures. This difference in explosive property determines the amount of rock liberated on detonation, together with the resulting end-form of the blasted face. The ability to determine how and where explosive charges are to be situated is another significant area of change in blast design and has resulted from the adoption of high explosives and the incorporation of the explosive slurry mixture of ammonium nitrate and fuel-oil. Modern quarry blasting uses these explosives in combination with the high-explosives shattering the lower part of a rock-face whilst the ANFO mixture heaves open the upper portion of the rock mass. Their use is complemented by time-delayed detonation of charges in the shot-holes. This differs greatly from the random heaving open of discontinuities which resulted from using blackpowder poured into prominent joints or packed into header tunnels at the foot of the quarry face. Little or no control could be exercised over the degree of fragmentation, size and position of blast piles or the resulting face angle, all of which have become much safer and more predictable in their outcome.

Adoption of more versatile explosives has been augmented by increased mechanization, efficiency and accuracy of rock drilling. The introduction of high pressure rotary and percussion drilling rigs capable of drilling more accurately orientated and diameter controlled shot-holes has enabled greater control to be exercised over the resulting end-form of the quarry face than was previously possible with earlier drilling methods and the random action of blackpowder. This has led to the excavation of more predictable end-forms for the quarry face following the detonation of explosive charges in commonly a series of shot-holes along a length of face to be blasted. The calculated distances between shot-holes and their position back from the existing quarry edge (burden) contrasts strongly with the blackpowder header blasts.

### Fracture Patterns

Rockfalls were found to be most frequent from three areas of the quarry face each with distinct patterns of rock fracturing. It was further apparent that significant widening of vertical and sub-vertical joints, together with distortion and realignment of bedding planes had also taken place. The first pattern of fractures occupying the upper third of the quarry faces is considerably more affected by blast, as evidenced by increased fracture density and bedding plane distortion, particularly near the crest of quarry faces. In contrast, the foot of quarry faces exhibit only limited bedding plane disturbance and much reduced fracturing. The positions of shot-holes are clearly evidence by regularly spaced, vertical white scorch marks along the quarry face. These are formed by the rapid expansion of explosive gases into the rock mass from the shot-holes. Between the shot-holes are areas possessing a high concentration of blast-induced fracturing, together with enhancement of existing natural discontinuities. There is complete disruption of bedding plane alignment in this part of quarry faces with a complex pattern of cross-bed fracturing producing an assortment of highly unstable and irregularly orientated blocks of limestone.

A second pattern of fractures is associated with rock buttresses. This consists of only limited bedding plane distortion but with a combination of increased vertical blast fractures and tension fracturing. The bases of rock buttresses possess prominent widened joints running vertically up the centre of the buttress in conjunction with similarly aligned vertical blast fractures. These can extend to over half the face height at their maximum extent towards the centre of the buttress but are reduced in both length, spacing and depth into the face towards the edges of buttresses. Large tension fractures occur parallel to the plane of the rock-face along the lateral edges of buttresses. This gives rise to a convex profile towards the foot of buttresses as columnar-shaped blocks, produced by widened joints and vertical blast fractures, slide down and out from the rock-face. Rockfalls are less frequent but of greater individual magnitude than from blast fracture cones.

The third pattern of fractures is radially orientated outwards from the centre of roughly circular scoops out of rock-faces. They occur beneath the apex of blast fracture cones and above the uppermost part of rock buttresses, occupying the middle portion of quarry faces.

Two further sets of discontinuities are present which are not related to the presence of a particular landform. The first of these occupies a position in the upper half of rock-faces immediately above the stemming-line, an arbitrary line marking the base of infilling material used to pack-down the explosives in each of the shot-holes. Bedding plane enlargement and realignment occurs together with cross-fractures which only rarely travel completely between beds. Below there is some bedding plane widening but no realignment or cross-fracturing. The final set of discontinuities occupy what may be considered as the least blast fractured parts of the quarry face. These areas lie at the foot of quarry faces between the outermost parts of rock buttresses. They exhibit substantial bedding plane widening but no realignment and very few cross-bed fractures.

## Landform development

Having considered these controls in relation to the quarry landform model it became clear that the position of certain landforms across worked faces accorded with the position and spacing of shot-holes as designated by the drilling and blasting design. Hence, debris collection traps were relocated beneath identified rock buttresses and blast fracture cones and rockfalls monitored from each for a selection of quarry faces. These observations demonstrated that the lateral development of blast fracture cones across the rock-face was limited by the presence of rock buttresses at either side. The extent of these buttresses varied with the age of the face, being more prominent across older faces and also over recently (less than two years) abandoned faces where blasting had excavated a face against the dip of the bedding planes. Blast fracture cones are found in various stages of collapse and generate differing magnitudes of rock falls. The size and shape of rocks found in the debris collection traps differed with elongate and predominantly wedge-shaped rocks falling from rock buttresses, whilst more angular and blocky-shaped rocks fell from blast fracture cones.

The pattern of shot-holes and their explosive charging are seen to control not only the nature of the production blast for which they are prepared but also the further development of the quarry face if blasting is discontinued. The alternating sequence of blast fracture cones and rock buttresses is found to be a characteristic landform sequence over those rock faces which employ a combination of ANFO and dynamite in the blasting design, accompanied by the use of stemming in each of the shot-holes fired. The regularity of these landforms across a length of excavated face and their accordance with the recorded position of shot-holes was repeated across faces which had been excavated using similar drilling and blasting designs (Figure 8.5). The blast fracture cones are located in the upper part of the quarry face above the stemming-line. No blast fracture cone was seen to develop beneath this line and all were restricted to the upper third of the total height of rock-face. These landforms are highly unstable and generate some of the earliest rockfalls from blasted faces on abandonment. This results in the widening of these cone-shaped areas of fractures across, down and back into the crest-line of the face. The apex of the cone rarely advances down beyond the stemming-line.

This can be explained by the sequence of events which occurs across the rock-face upon the detonation of the explosive charges. The aim of the drilling and blasting design is to excavate safely the maximum amount of stone of the desired fragmentation into an easily removed blast pile, whilst leaving the rock-face in a condition suitable for further blasts to take place. The shot-holes are charged

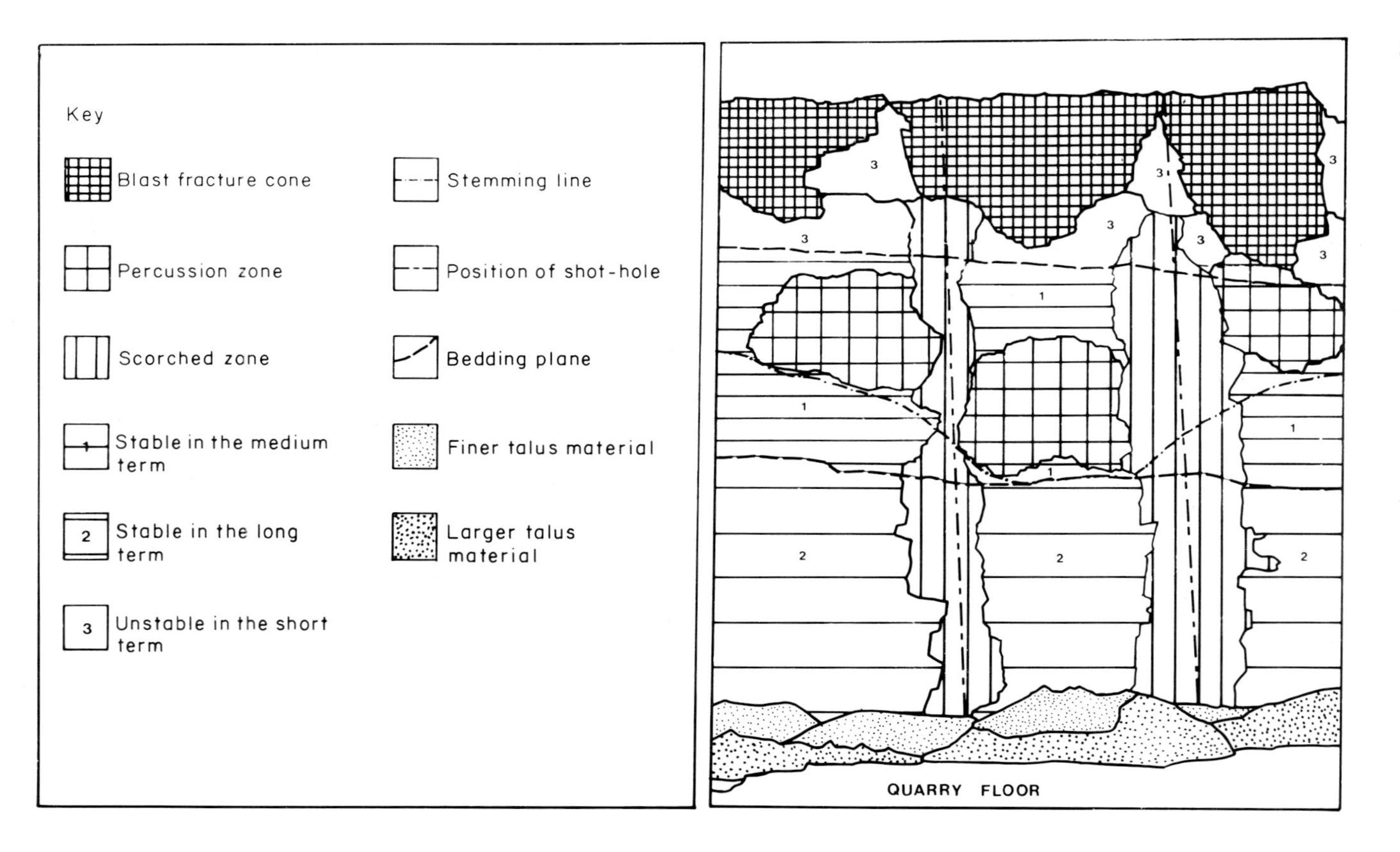


**Figure 8.5** Diagram of a section of blasted quarry face

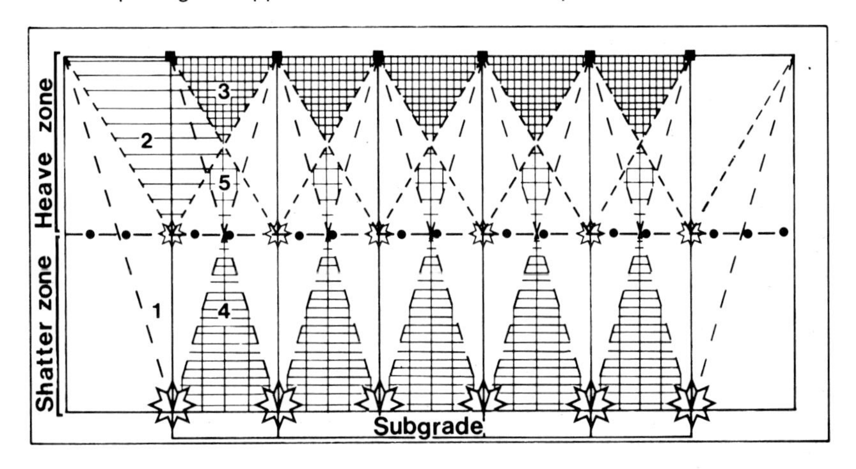


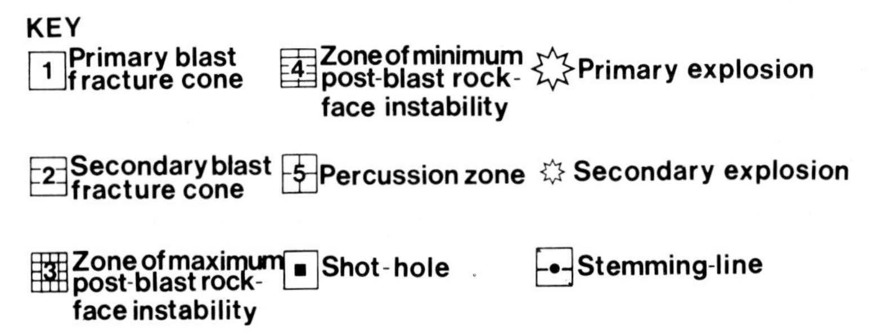


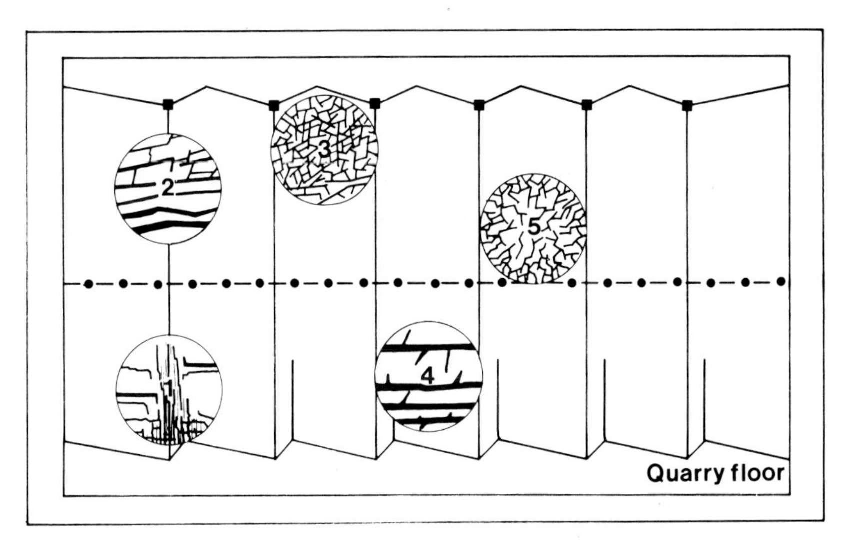


**Figure 8.6** Blast fracture pattern initiation

so as to excavate the burden out from the line of the face onto the quarry floor. This is effected by the delayed detonation of the high-explosive at the base of each of the shot-holes which have been drilled down into the subgrade. This reduces the likelihood of a rock stump remaining in the core of the blast pile which would hinder mechanical shovel clearance of stone. The ANFO placed higher up shot-holes is detonated almost immediately after these base charges. The aim is to shatter effectively the lower half of the quarry face with the high-explosive and to push the rock out across the quarry floor. The limited time delay of the 'heaved' portion of the quarry face above aims to lift the rock upwards and outwards allowing it to fall down on top of the earlier shattered rock beneath.

The detonation of these explosives charges in each of the shot-holes produces a primary cone of blast fracturing from the high-explosives, and a secondary cone of blast fracturing from the ANFO detonation which it augments. These overlap each other across the quarry face combining to produce alternate areas of intensified blast effects (Figure 8.6). It is the foci of these detonations which establishes the complex pattern of fracturing and bedding plane realignment described for blast fracture cones. The intense shattering force which accompanies the detonation of the high-explosive at the base of each of the shot-holes is responsible for the vertical fracturing and joint widening which characterizes rock buttresses. Seepage of ANFO slurry into fractures emanating out into the burden from the shot-holes further enhances the percussive effect of this secondary detonation between the shot-holes, resulting in the shallow scoops out of rock-faces (percussion zones). Rock buttresses remain as prominent features projecting out from quarry faces because most of the explosive force travels up and, importantly, out from the shot-hole in the first instant of detonation. However, as the explosive force travels up the shot-hole it is able to expand outwards into the burden and, augmented by the ANFO detonation, funnels-out as it continues upwards. This results in a repeated, near circular fracturing pattern at the crest of rock-faces. This subsequently half-collapses back down onto the blast pile beneath leaving only a semi-circle of fractures which will subsequently collapse to form the blast fracture cones.

## RESTORATION BLASTING

The ability to predict the likely future development of a blasted rock face by the identification of areas on the face which will be more or less stable, can be applied to the adoption of new drilling and blasting designs aimed at the restoration of the quarry face. Restoration blasting is the application of a series of drilling and blasting designs in order to reduce the engineered appearance of a production blasted quarry face. The overall aim is the formation of a daleside landform sequence through the construction of skeletal rock landforms

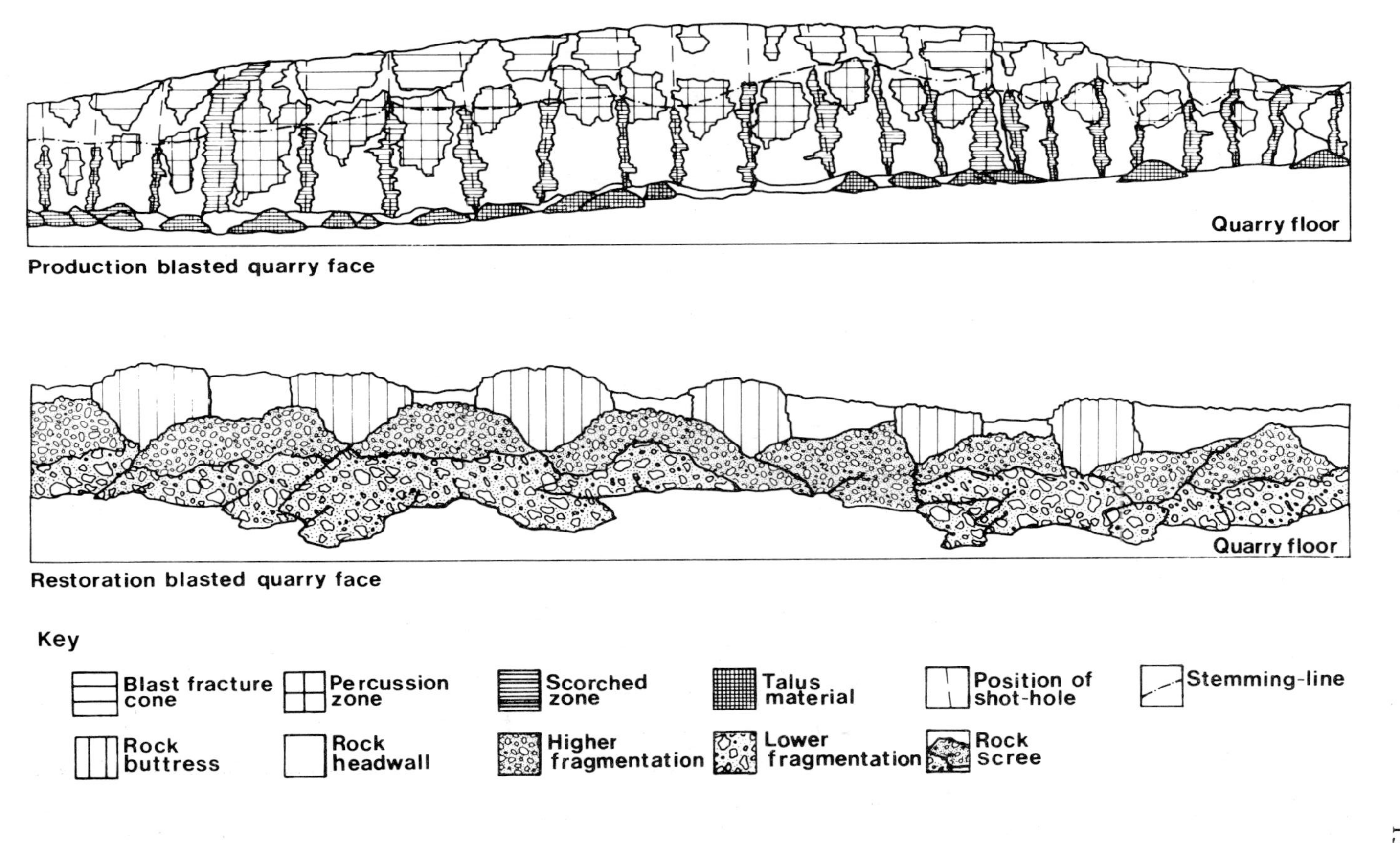


**Figure 8.7** A potential daleside landform sequence, Section-6D, ICI Tunstead Quarry, Buxton

consisting of rock headwalls, buttresses and screes, the scale and extent of which will mimic those of a natural limestone daleside (Figure 8.7). The construction of these skeletal landforms together with their subsequent revegetation will enable quarried rock faces to be more easily harmonized with the surrounding unexcavated landscape. The technique has two main elements which are the subject of ongoing research, the construction of skeletal rock landforms and their infilling and colonization.

### Construction of Skeletal Rock Landforms

Four specific objectives have been identified:

(1) Reduction of face height by the construction of scree blast piles which will:

- (i) Mask the regular sequence of scorch marks of previous production blasts.
- (ii) Cover the quarry face to varying heights thereby reducing the extent of face available to liberate rock falls.
- (iii) Have varying angles of rest to differ from the relative uniformity of production blast piles.
- (iv) Vary in their degree of fragmentation both vertically and laterally.

(2) Indentation of the crest line by a series of semicircular cut-backs to mimic the collapse of natural sinkholes and blast fracture cones.

(3) Formation of a 'ragged' rock headwall in the upper third of the quarry face.

(4) Stabilization of the scree blast piles by leaving a rock stump at the base of the face and/or varying the fragmentation of the blasted rock down through the scree blast pile with coarser material at the base.

### Infilling and Colonization of Skeletal Rock Landforms

Following the construction of these skeletal rock landforms it will be necessary to identify suitable infilling materials for the scree slopes with regard to their stability and vegetation colonization. An evaluation of the potential of quarry waste materials and stripped overburden will precede their application and the establishment of vegetation trials. Investigations of the form, hydrology and long-term stability of the vegetated landform sequence will be undertaken to establish the success of restoration.

## POLICY AND PLANNING IMPLICATIONS

Prior to the 1940s limestone was quarried by hand methods in small, locally owned and operated quarries. There were few environmental conflicts between the interests of the quarrying companies and their local communities. Mineral working was a way of life often undertaken alongside hill farming as a major part of the local economy. Quarrying companies had substantial areas of land within their control, finding it easy to obtain working rights with minimum royalties. No planning consent was needed and the scale of working was easily accommodated into the landscape once working ceased. The advent of increasing demand led to rapid technical and mechanical developments in the extraction of limestone with a corresponding increase in the size of workings. Many quarrying companies consolidated their long-term position during the 1940s by gaining extensive mineral rights when few planning conditions were attached to the consents granted. There was no legal obligation to restore these workings beyond, in some cases, the removal of redundant plant and machinery. The advent of the Town and County Planning Act in 1947 set in motion increasingly stringent requirements for mineral operators to take account of the impact of their quarrying operations.

The early 1970s saw a rapid rise in demand for aggregates which led to increased concern as to how this demand was to be met. The Verney Committee on Aggregates (Verney, 1976) was established to rationalize demand with regional sources of supply. The Committee was particularly conscious of the need for mineral operators and planning authorities to work together at regional level to achieve an improved understanding of the issues involved, including the winning of materials in environmentally sensitive areas. The development of planning controls over mineral workings is the principal vehicle by which the Government, county and local councils control those aspects of the minerals industry thought likely to have an adverse effect on the environment. The impact of quarry workings varies greatly and is related not only to the area excavated but its relationship with its surroundings. In particular the engineered appearance of blasted quarry faces and the scale and extent of the resulting quarry rock basin may contrast strongly with the surrounding unexcavated landscape.

The early 1970s were also a period of increased environmental awareness and as a result those responsible for developing minerals policy, notably county council planning authorities, began to examine critically the activities of the extractive industries. Particular concern was expressed over land dereliction resulting from mineral working and this led to the establishment in 1972 of a Government committee, under the chairmanship of Sir Roger Stevens, with instructions to examine the operation of statutory provisions under which planning control was exercised over mineral workings. The committee documented its findings in the report *Planning Controls over Mineral Working* (Stevens, 1976). Public concern and examination of their activities caused minerals operators to respond with a vigorous defence of their operations including the publication of the Zuckerman report in 1972. Representations were made to local planning authorities regarding the need for unobstructed long-term planning and certainty of development permissions in order to ensure the economic viability of mineral workings.

The late 1970s saw a recognition of the need to improve minerals planning legislation in order to alleviate the environmental impact of the minerals industry. Minerals policy was largely aimed at containing the detrimental effects of mineral working and included controls on air and water pollution, blast and processing noise, vibration, subsidence and waste utilization. The need for rehabilitation, however, whilst receiving widespread recognition was not as equally well-supported. Whilst consideration has been given to sand and gravel and aggregate waste materials less attention has been focused upon crushed rock. Indeed recommendations for the rehabilitation of hard rock quarries have drawn heavily upon practices for sand and gravel and open-cast coal working which are often inappropriate in both scale and detail. The DoE report 'The Environmental Impact of Large Stone Quarries and Open-Pit Non-Ferrous Metal Mines' (Down and Stocks, 1976) looked at the nature of hard rock quarries and described a number of environmental impacts associated with the extraction, processing and transportation of quarried aggregates. It identified wide-ranging research needs including landscaping, revegetation and considerations for after-use.

The minerals industry has traditionally sought to find a compromise between responding to environmentally detrimental aspects of its operations and the economic commitments of the market place. Planning authorities have striven to become more aware of the nature of extraction methods and to acquaint themselves with the intricacies of the minerals market locally and nationally. This is a process of evolution with attitudes changing, and being changed, by individual circumstances. The minerals industry seeks to improve its response to planners, together with its image to the wider public, and to secure permission to expand and develop its operations with an environmentally aware outlook. Minerals planners aim to alleviate the impact of minerals extraction, processing and wastes disposal during the working life of the site whilst securing satisfactory arrangements for its immediate and longer term after-care and use.

Current policy considerations are increasingly being focused upon the need for suitable rehabilitation programmes to be implemented as an on-going process in the development of the mineral working with restoration plans being determined, approved and regularly reviewed between the mineral operator and the planning authority. Since 1981 minerals planning authorities have had much greater control on the development of existing operations and on the establishment of new workings through the Town and Country Planning (Minerals) Act 1981. However, policies for the practical and economic restoration of hard rock quarry faces have fallen behind restoration guidelines and planning constraints applied to both open-cast coal and sand and gravel workings. This is in part a consequence of the necessity for quarry workings to be economically viable for at least fifty years making the development and implementation of a rehabilitation strategy an extended process often beyond the working lives of those who first define its intentions.

The determination of a model of limestone quarry evolution, together with the development of restoration blasting, will provide minerals operators, planners and legislators with an opportunity to incorporate the reconstruction of natural landforms into quarry rehabilitation programmes. The production of visually attractive, safer and more predictable landforms by the application of restoration blasting will enhance the success of current restoration practices and increase the diversity of possible after-uses for abandoned limestone quarries. This will enable quarry restoration programmes to incorporate worked faces more harmoniously into the surrounding landscape.

## CONCLUSION

The research described in this chapter was initiated to gain an understanding of the geomorphological impact of limestone quarrying and in particular the post-abandonment evolution of quarried limestone rock slopes under the influence of natural processes. It has described landform development on quarried limestone rock slopes in a group of twelve quarries in the White Peak, Derbyshire. A model of quarry landforms has been described and used in the interpretation of how differing methods of rock slope excavation, particularly the drilling and blasting design employed, have affected the development of quarried limestone rock slopes once active working ceases. The ability to predict the likely future development of a blasted rock-face, by the identification of areas on the face which will be more or less stable, can be applied to new drilling and blasting designs aimed at the restoration of quarried limestone rock-faces. The formation of daleside landform sequences by the construction of skeletal rock landforms which not only mimic the outward form of their natural counterparts but can be predicted to evolve in harmony with the operation of natural processes, is the aim of restoration blasting. Three restoration blasting trials have been undertaken in Tunstead Quarry, Buxton by arrangement with ICI plc and these produced landforms which possess many of the characteristics required to both ensure their stability and, following the application of suitable infilling materials, to support vegetation. Future research will involve further restoration blasting trials in Tunstead together with the determination of infilling materials suitable for colonization by vegetation. A programme of revegetation will then be initiated and the characteristics of the constructed landforms will be investigated and their subsequent evolution monitored. It is also hoped that research will be undertaken in other limestone areas in Britain and overseas in order to assess the applicability of restoration blasting theory in working and abandoned limestone quarries in a range of geological settings. Ultimately it is intended that this geomorphological approach will provide practical and economic techniques which are acceptable to mineral operators, planning authorities and the wider public. These will then form part of wider rehabilitation strategies and mineral planning policies which enable the extraction of essential minerals to be undertaken but ensure that its impact upon the landscape is minimized.

### Acknowledgements

The investigation was made possible by provision of a Research Assistantship from the Faculty of Community Studies, Manchester Polytechnic. We would also like to thank ICI plc and the Peak Park Joint Planning Board for their technical and financial assistance with field work.

## REFERENCES

Dearden, J. (1963). Derbyshire limestone: its removal by man and by nature, *East Midlands Geogr.*, **3**(4), 199–205.

Down, C. G. and J. Stocks (1976). The environmental impact of large stone quarries and open pit non-ferrous metal mines in Britain, *Research Report 21*, Department of the Environment.

Gagen, P. J. and J. Gunn (1987). A geomorphological approach to restoration blasting in limestone quarries. In B. F. Beck and W. L. Wilson (eds) *Karst Hydrogeology: Engineering and Environmental Applications*, pp. 457–62, A. A Balkema.

Gunn, J. (1985). Pennine karst areas and their Quaternary history. In R. J. Johnson (ed.) *The Geomorphology of North-west England*, pp. 263–81, Manchester University Press.

Gunn, J. and P. J. Gagen, (1987). Limestone quarrying and sinkhole development in the English Peak District. In B. F. Beck and W. L. Wilson (eds) *Karst Hydrogeology: Engineering and Environmental Applications*, pp. 121–6, A. A. Balkema.

Gunn, J., D. Hardman, and W. Lindesey (1985). Problems of limestones quarrying in and adjacent to the Peak District National Park, *Annales de la Société Geologique de Belgique*, **108**, 59–63.

Gunn, J., P. J. Gagen and D. Raper (in prep.). Limestone quarrying and karst erosion rates in the English Peak District.

Haigh, M. J. (1978). Evolution of slopes on artificial landforms, Blaenavon, U.K., *Geogr. Res. Pap. 183*, University of Chicago Press, 293pp.

Harrison, D. J. (1981). *The Limestone and Dolomite Resources of the Country Around Buxton, Description of 1:25000 Sheet SK 07 and Parts of SK 06 and 08*, Mineral Assessment Report 77, Institute of Geological Sciences, HMSO, London, 108pp.

Sherlock, R. L. (1922). *Man as a Geological Agent—An Account of his Action on Inanimate Nature*, London.

Stevens, Sir Roger, Fleming, M. G., Nardecchia, T. J. and Taylor, J. C. (1976). *Planning Control over Mineral Working*, Report of the Committee under the chairmanship of Sir Roger Stevens GCMG, HMSO, 448pp.

Thornes, J. B. and Brunsden, D. (1977). *Geomorphology & Time*, Methuen & Co. 280pp.

Verney, R. (1976). *Aggregates: The Way Ahead*, Report of the Advisory Committee on Aggregates, HMSO, London. 118pp.

Zuckerman, Lord., Arbuthnott, Viscount of, Kidson, C., Nicholson, E. M., Warner, Sir F., and Longland, Sir J. (1972). Report of the Commission on Mining and the Environment, London. 92pp.

# River Management

Geomorphology in Environmental Planning
Edited by J. M. Hooke



# 9 Channelization, River Engineering and Geomorphology

**ANDREW BROOKES**
*Thames Water, Nugent House, Vastern Road, Reading*

**KEN GREGORY**
*Department of Geography, University of Southampton*

Channelization is the term used to embrace all processes of river channel engineering for the purposes of flood control, drainage improvement, maintenance of navigation, or reduction of bank erosion (Keller, 1976; Brookes, 1985a). Channelization may be undertaken by engineering procedures which either enlarge, straighten, embank or protect the existing channel, or which involve the creation of new channels. Other channelization procedures may be classified as river channel maintenance, and include dredging, cutting or the removal of obstructions.

Geomorphology largely ignored the implications of river engineering works until the mid-twentieth century but since the late 1960s has utilized greater understanding of the interrelations of river channel form and process as the foundation for studies of river channel changes (Gregory, 1977). Research has subsequently focused upon the obvious changes of river channels which occur at sites of engineering works but has then been extended to the less obvious downstream effects. Whereas research on river channel changes first concentrated upon channel metamorphosis which provides applicable results, this has subsequently led to concern with aspects of river channel design and hence with research that is more strictly applied (Gregory, 1979). Channelization effects have prompted a considerable amount of research in view of the serious effects that channelization schemes produced, especially in the United States. Against this background the contributions made by geomorphologists to public policy on river engineering and channel management can be envisaged as including analyses of the distribution and extent of channel works; interpretation of the effects that are known to result from channelization; and the formulation of alternatives which have recently been developed as a means of working more sympathetically with the natural environment. The contributions which can be made by fluvial geomorphologists are influenced by the legislation requirements of individual countries and by the policies of individual organizations responsible for river management. Considerable scope still remains for further advances to be made in the applications of geomorphology, including the development of alternative strategies of channelization which are basin specific.

Such alternative strategies have emerged as we have proceeded from the approach portrayed by Leopold (1977) as one dominated by the notion that 'technology can fix it' to a much softer sympathetic treatment which has been characterized by Winkley (1972) and others as 'working with the river rather than against it'.

For full use to be made of the results of geomorphological research in relation to public policy and for further geomorphological contributions to be developed, it is essential that the extent of channelization, and the effects that works have upon river channels are appreciated by policy makers as well as the need to consider alternative designs and to manage the river channels in the context of the drainage basin unit. This chapter therefore proceeds by considering the extent of channel works, proceeds to summarize the consequences of channelization and then to show how prediction of channel recovery can be estimated, prior to outlining alternatives currently available in terms of legislation and policy constraints and then arguing for a more drainage basin approach to be adopted by policy makers.

## EXTENT OF CHANNEL WORKS

It is necessary to know the spatial extent of river channelization and over the last two decades considerable attention has been given to the character and distribution of channelization in the United States of America, where Leopold (1977) indicated that 26 550 km of major works had been undertaken and a further 16 090 km were then proposed. The specific character and extent of channelization in England and Wales has been acknowledged more recently, but the density of channelized river in England and Wales averages 0.06 *km* km$^{2}$, compared with a density of 0.003 *km* km$^{2}$ in the USA (Brookes, Gregory and Dawson, 1983). There is therefore a density of channelized river in the UK which averages 20 times the density of channelized river in the USA. In a national survey of channelization work undertaken in the fifty year period from 1930 to 1980, the national distribution was recorded in two main categories (Figure 9.1): capital works and major improvement schemes, shown as thick lines, and rivers which are maintained and are designated main river which are shown as lighter dotted lines. The proportion of these two categories of modification varies over England and Wales (Figure 9.2). The first category includes those types of channelization which may be regarded as having a major and lasting impact on channel morphology such as embanking and embankment improvement, channel enlargement by widening or deepening, the realignment of channels, bank protection and the lining of channels with concrete or

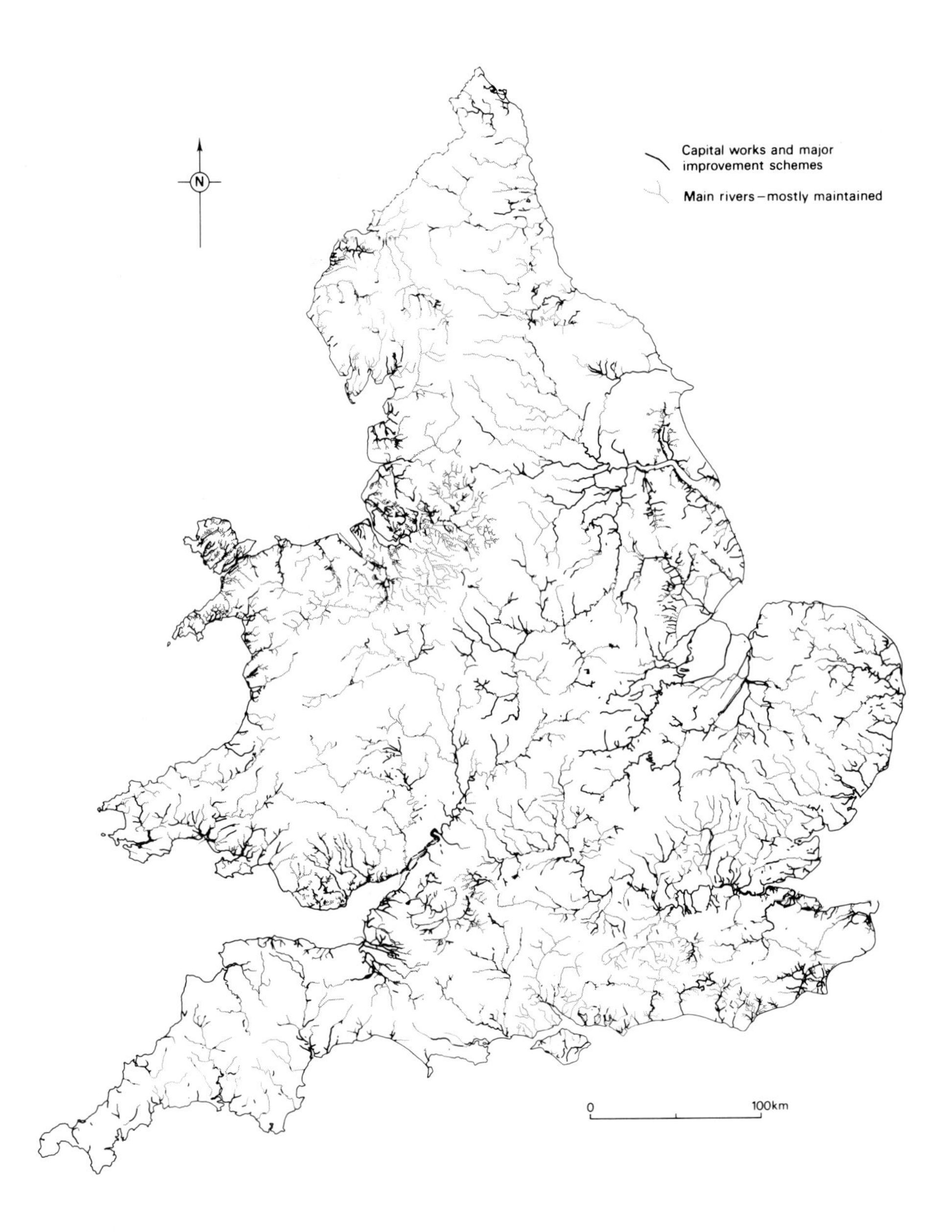


**Figure 9.1** Rivers channelized in England and Wales 1930–80. (Reproduced from Brookes, Gregory and Dawson, by permission of Elsevier Science Publishers B.V., Amsterdam. 1983)

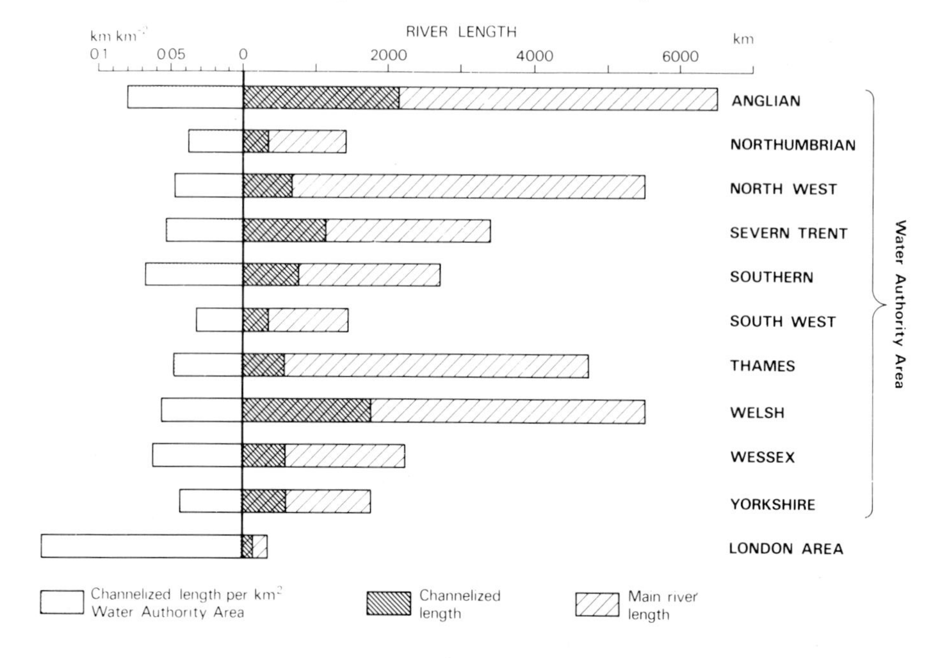


**Figure 9.2** Length of channelized river in water authority areas of England and Wales. Schemes included were those undertaken 1930–80 by Catchment Boards, River Boards. River Authorities, and Regional Water Authorities (from Brookes, Gregory and Dawson, 1983)

similar materials. This category is shown by the dense shading (Figure 9.2) and no distinction is made between channelization of one or both banks. The second category, of main river, includes those reaches for which individual organizations are responsible for maintenance. Such maintenance has frequently been undertaken on a routine basis over the period, has often involved considerable lengths of river channel, and has sometimes occurred as much as two or three times per year. Maintenance activities include weed control, the removal of bankside vegetation, dredging of accumulated shoals, and the clearance of rubbish from urban channels.

Over the Water Authority areas of England and Wales there is some considerable variation in the percentage of main river which is channelized (Figure 9.2) and this varies from 41 per cent in the London area and 34 per cent in the Severn, Trent and Yorkshire Water Authority areas, to 12 per cent in the North West area.

Such surveys of the distribution and character of channelization are necessary as a basis for considering the effects that channelization has had and the alternative ways in which the objectives can be achieved. However, when it is recalled that some 8500 km of river in England and Wales have been subjected to major or capital works, and that a further 35 500 km of river is maintained as main river, this underlines the significance of channelization and emphasizes why it must be considered in detail. It is also important to remember that there has been a very considerable amount of channelization along the rivers of England and Wales prior to the pattern representative of the period 1930–80 (Figure 9.1). Thus in England and Wales and elsewhere in the world it is not always easy to establish the magnitude, extent and frequency of channelization work that has been undertaken in the past.

## REPERCUSSIONS FROM CHANNELIZATION

The dramatic and adverse effects that channelization practices can have were first appreciated in the United States. Since channelization involves manipulation of one or more of the dependent hydraulic variables of slope, depth, width and roughness, then feedback effects may be instituted which proceed to promote a new state of equilibrium. Channelization projects, particularly if their design is based on inadequate background data, may induce instability not only in the improved channel reach but also upstream and/or downstream of the reach. In each section erosion or deposition may take place. Perhaps the most dramatic adjustments occur in response to slope changes associated with channel straightening or regrading, or to extended bottom widths.

Examples of the morphological consequences of artificial straightening are summarized in Table 9.1. Straightening meandering channels increases the river channel slope because of the shorter channel path. An increase of slope enables the transport of more sediment than is supplied by the natural channel upstream, such that the additional sediment supply is obtained from the bed of alluvial channels by degradation which progresses upstream as a nickpoint (Parker and Andres, 1976). The excess load arising from the erosion is then transported to the downstream portion of the straightened reach, and because of the flatter natural reach downstream, the sediment is deposited on the bed. In the 38 years following straightening of the Willow River in Iowa erosion increased the capacity by 440 per cent (Daniels, 1960), whilst in Missouri the cross-sectional area of the Blackwater River increased by a maximum of 1000 per cent over a 60-year period (Emerson, 1971). Rapid adjustments may have significant engineering consequences, and straightening of the Lang Lang River in Australia between 1920–23 caused severe degradation which destroyed a total of seven bridges (Bird, 1980).

Results from specific research investigations (Table 9.1) afford some indication of the morphological consequences of channelization reported within the area of the scheme, in the reaches upstream and downstream and also of the spatial extent of the changes. In addition it is necessary to gain some idea of the range of effects of channelization on a world scale and of how these effects compare with other river channel changes. To provide such a background for the effects

**Table 9.1** Examples of the morphological consequences of river channelization

| Area applied | Dates | Extent | Morphological effects | | | Source |
|---|---|---|---|---|---|---|
| | | | Downstream | Within | Upstream | |
| Willow R, Iowa | 1906–20 | | | 440 per cent capacity increase between 1920 and 1958 | Series of nick-points | Daniels, 1960; Ruhe, 1970 |
| Blackwater R, Missouri | 1910 | | | 1000 per cent increase of cross-sectional area | | Emerson, 1971 |
| East and West Prairie rivers, Canada | 1953–71 | | Deposition of sediment | 500 per cent increase of cross-sectional area | 4 metres of degradation (1964–74) | Parker and Andres, 1976 |
| Tisza R, Hungary | 1850s | 32 per cent reduction of length | No effect—it enters the Danube | Bed lowered by a maximum of 2.3 metres | | Szilagy, 1932 |
| Mississippi R, Baton Rouge | 1929–47 | 35 per cent reduction of length | No effect due to lack of time since completion | | | Lane, 1947 |
| Colorado R, Laguna Dam | | 9.6 km length reduction | Insufficient sediment to raise bed level | Bed lowered | Lowering limited by dam located upstream | Lane, 1947 |
| Peabody R, New Hampshire | | | | Reach lowered by a maximum of 3.6 metres; width increase of 400 per cent | | Yearke, 1971 |
| 103 streams in North America | 1960–70 | Lengths of 0.07 to 4.2 km | | Degradation at only 17 sites; bank stability poor at 7 sites | | Brice, 1981 |

that channelization can have upon the morphological characteristics of river channels, 182 published studies of river channel change including channelization have been investigated. Figure 9.3(a) and (b) depicts five types of change due to reservoir and dam construction, land use changes, urban development, point changes and water transfers. In each case the vertical line represents the situation of no change. Each study is represented by a horizontal line and deviations are shown to the left if there has been a reduction in channel dimensions and to the right if there has been an increase in channel dimensions. The wide range of results demonstrates that adjustments of river channels have taken place on a range of scales for a number of reasons and that substantial adjustments have occurred involving increases or decreases by factors of up to 10 times. Results from studies investigating schemes which are exclusively channelization are shown in Figure 9.3(b) as heavy lines and these are usually characterized by increases in channel dimensions.

Channelization procedures are not always confined to a reach or small section of a river channel and to the immediate downstream area, because they often may also be associated with catchment land-use changes. For example, in the land use category (Figure 9.3(a)) line 9 represents a reclamation programme involving dikes, ditches and river cutoffs in the Pemberton Valley, British Columbia, Canada. Similarly urban development (Figure 9.3(b)) often includes channelization procedures applied to urban river channels as well as more extensive impervious areas. In the case of areas affected by urban development, it is evident that channel dimensions can increase by as much as 10 times and the vertical shading on the diagrams (Figures 9.3(a),(b)) relates to increases of 1, 3, 5, 7 and 9 times. Most of the river channel adjustments arising from other point changes (Figure 9.3(b)) include elements of channelization procedures and so they further demonstrate the effects that channelization has on river channels.

The results of research investigations from a number of areas in the world (Figures 9.3(a),(b)) emphasize how much change has occurred and how diverse the changes can be. The studies included in the diagrams embrace a range of channel adjustments including changes of channel cross-section (capacity, width, depth, hydraulic radius and width–depth ratio), and of channel reaches (volume, sinuosity, meander wavelength, channel slope, channel length, channel island area, surface area and floodplain area). The particular combination of parameters or degrees of freedom which adjust in any particular area is a response to drainage basin and local channel controls. Although it is not easy to predict what will change at any particular location, the further adoption of a basin approach (as indicated below) will assist understanding and facilitate prediction.

Many channels adjacent to channelization schemes are enlarged as a response to process changes induced by channelization. However, without maintenance, channels deliberately over-widened may revert to their original dimensions because reduced velocities at low flows cause sediment to be deposited.

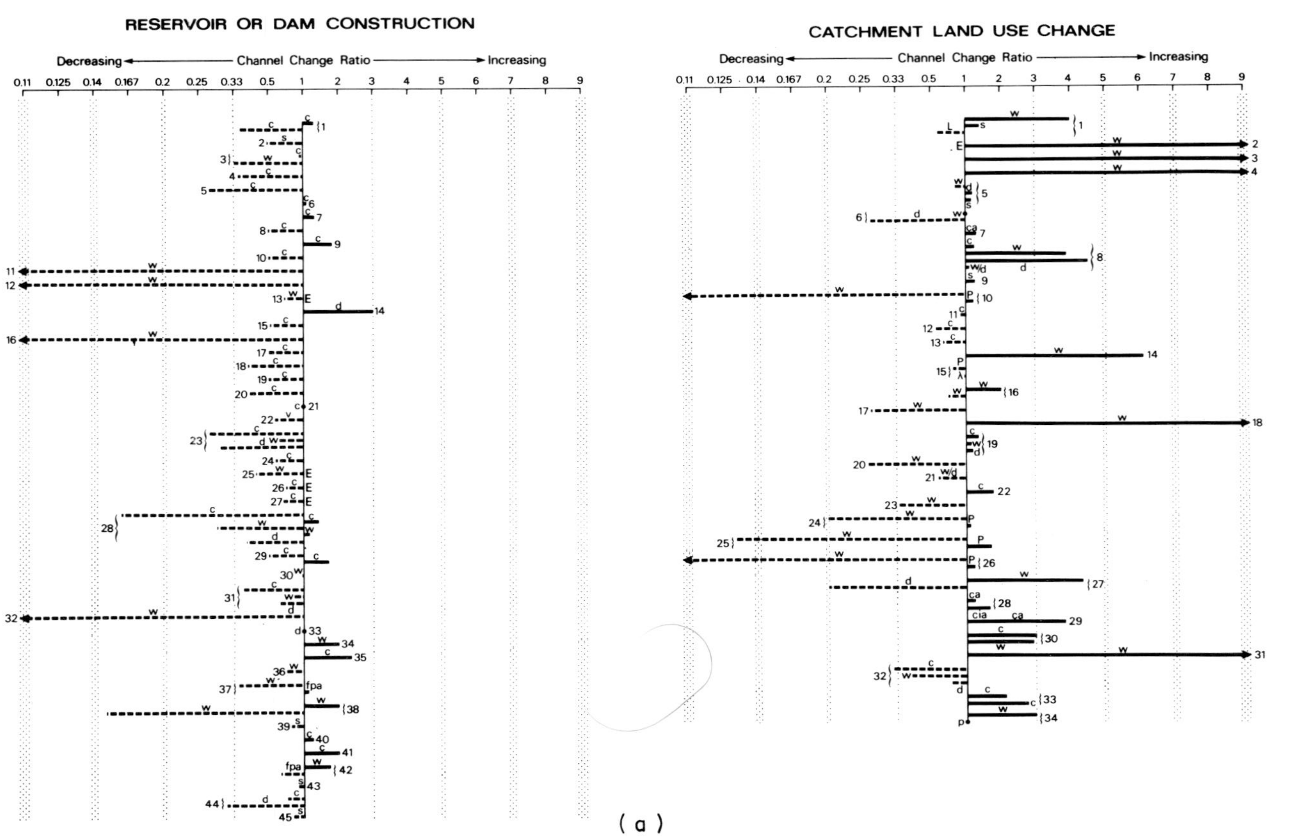


**Figure 9.3** Nature and magnitude of river channel changes according to published research results for a variety of world areas. Studies are classified according to effects of reservoir or dam construction and to catchment land use change in (a) and according to urban development, other point changes, and water transfers in (b). In (b) the point changes which are exclusively channelization are shown in heavy lines. The key to nature of channel adjustment is given in (b)

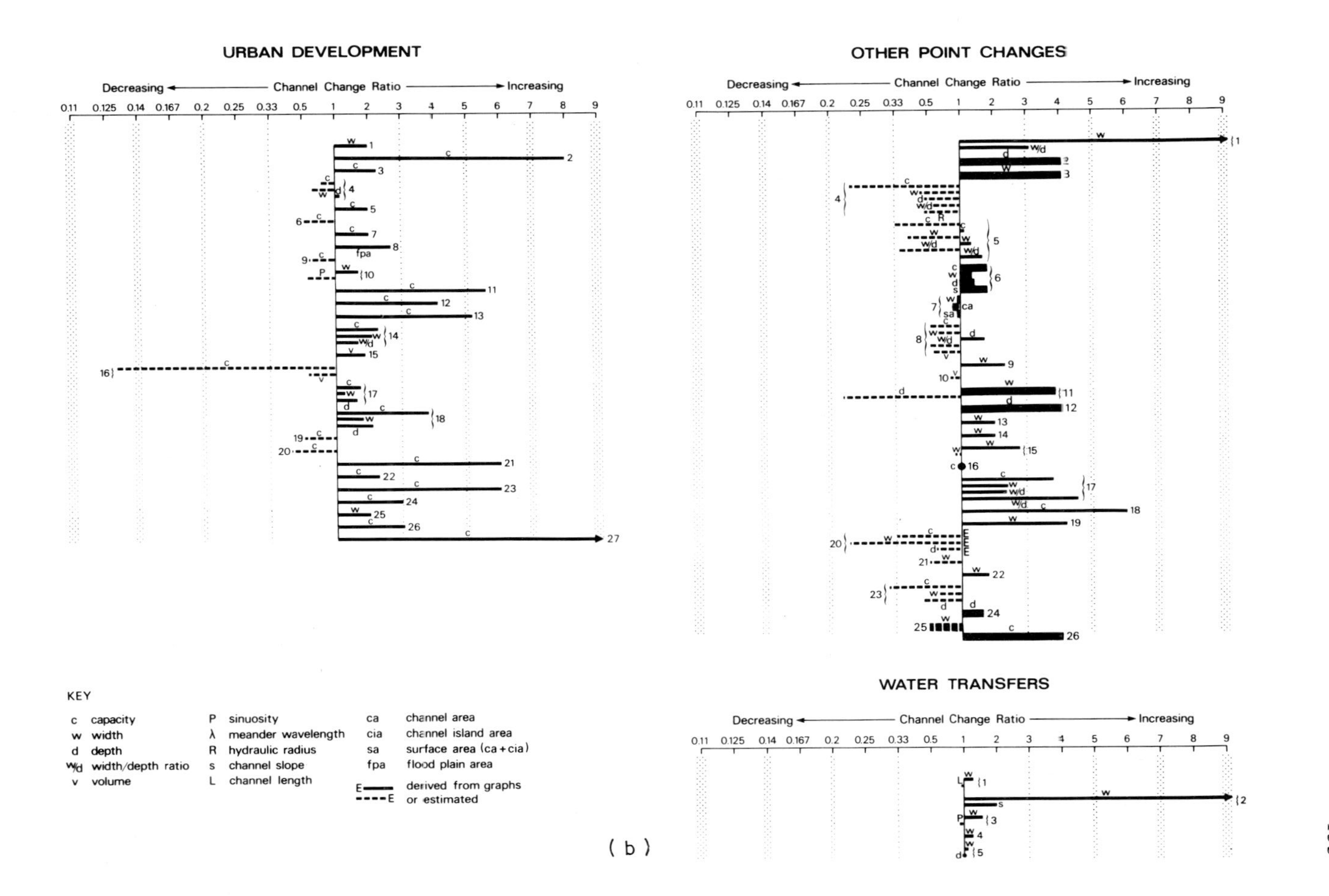
URBAN DEVELOPMENT
Decreasing ← Channel Change Ratio → Increasing
0.11 0.125 0.14 0.167 0.2 0.25 0.33 0.5 1 2 3 4 5 6 7 8 9
OTHER POINT CHANGES
Decreasing ← Channel Change Ratio → Increasing
0.11 0.125 0.14 0.167 0.2 0.25 0.33 0.5 1 2 3 4 5 6 7 8 9
WATER TRANSFERS
Decreasing ← Channel Change Ratio → Increasing
0.11 0.125 0.14 0.167 0.2 0.25 0.33 0.5 1 2 3 4 5 6 7 8 9
KEY
c capacity
w width
d depth
w/d width/depth ratio
v volume
P sinuosity
λ meander wavelength
R hydraulic radius
s channel slope
L channel length
ca channel area
cia channel island area
sa surface area (ca + cia)
fpa flood plain area
E derived from graphs
E or estimated

(b)

Such deposits may become stabilized by vegetation to form more permanent benches which cannot be mobilized during flood flows (Nixon, 1966). Reaches which have been deepened may serve as sediment traps, such that flow depths and channel capacities are progressively reduced. Within fifteen years a flood channel on the San Lorenzo River in California had been infilled by 350 000 $m^3$ of sediment, the capacity being reduced from the 100-year design flood to about the 25–35-year flood (Griggs and Paris, 1982).

This indicates that there are temporal changes following any channelization scheme and such adjustments or readjustments over time may feature within the scheme itself or in adjacent reaches. Although the downstream consequences of channelization have received insufficient attention for spatially contiguous areas, a study of 57 sites in England and Wales revealed the nature and extent of adjustments which have occurred below a variety of types of channelization (Brookes, 1985b; 1987a). The majority of sites located in high energy environments in upland Britain had downstream reaches which had clear indications of erosion and width had changed preferentially to depth at these sites. It was possible to produce a multiple regression equation (Brookes, 1985b) to indicate the amount of adjustment downstream of channelization schemes in England and Wales and one of the variables included was age of scheme thus indicating the significance of the time factor.

Manipulation of hydraulic variables by channelization, and any subsequent adjustments of the channel morphology may produce corresponding dramatic consequences for the aquatic ecosystem. Loss of morphologic variability and erosion and sedimentation have substantial effects on all forms of aquatic life. Perhaps the best documented ecological effects are those relating to fish and fisheries (Table 9.2). A commonly stated reason for change in fish populations is the loss of a natural pool-riffle sequence which provides a variety of low flow conditions suitable as cover for fish and for the organisms on which fish feed. Shelter areas are required at high flows to protect fish from abnormally high water velocities and these conditions are absent where a meandering stream is replaced by an artificially straightened channel composed mainly of riffle situations. Manipulation of the width and depth variables in a channel may create shallow and unnatural flows which are an unsuitable habitat for fish and may present topographical difficulties for fish migration (Bayless and Smith, 1967; Graham, 1975; Headrick, 1976). Disruption of the substrate may also affect bottom dwelling organisms which are an important source of food for fish (Dodge *et al.*, 1977; Griswold, 1978). Furthermore, clearance of bankside and instream vegetation during channelization may destroy valuable cover for fish.

Channelization has thus led to the modification of habitats downstream of channelization schemes and often to the complete obliteration of habitats within the scheme itself. Whereas channelization schemes have been successful in achieving the primary objectives of controlling flooding and erosion and improving drainage and facilitating navigation, more recent research has revealed

**Table 9.2** Examples of the ecological impacts of channelization

| Location | Type | Date | Variables affected | Reduced parameters | Recovery | Source |
|---|---|---|---|---|---|---|
| Missouri R, Nebraska | | 1950s | Brush and piles and pools eliminated | Benthic area reduced by 67 per cent; standing crop of drift reduced 88 per cent | Study represents 15 years after channelization | Morris *et al.*, 1968 |
| Little Sioux R, Iowa | | | Lack of suitable substrate | Fewer macroinvertebrates; but more drift organisms | | Hansen and Muncy, 1971 |
| Buena Vista Marsh, Portage Co, Wisconsin | Dredging | Various | Seasonal effect; substrate only unstable at high flow; at other times vegetation/silt | High invertebrate populations when substrate stable; vegetation and silt favours snails/midges. Elimination of stoneflies. | | Schmal and Sanders, 1978 |
| Mud Creek, Douglas Creek, Kansas | Enlarged | 1971 | Destruction of pool-riffle sequence | Number of fish and species reduced; biomass reduced by 82 per cent and diversity by 49 per cent | Two years after | Huggins and Moss, 1974 |
| Rush Creek, Modoc County, California | Realigned | 1969 | Destruction of pool-riffle sequence | Trout biomass reduced by 86 per cent; total number by 36 per cent and total biomass by 69 per cent | Five years after construction | Moyle, 1976 |
| Missouri River | Various channel works | | Lack of niches (especially pools) | Annual catch reduced by 47 per cent; harvest rate by 48 per cent; and standing crop by 63 per cent | | Groen and Schmulbach, 1978 |

how serious morphological and ecological consequences can arise. In the USA the Committee on Government Operations (1973) reviewed much of the evidence associated with channelization effects and it is from such studies that it has been necessary to consider prediction and recovery (next section) and then to devise alternative practices (following section).

## PREDICTION AND RECOVERY

Unfortunately without a complete set of deterministic equations it is not possible to predict precisely the morphological response to alterations of width, depth, slope, roughness or channel planform caused by channelization. However, channel behaviour can be deduced to a certain extent by integrating responses of stream channels that have already been altered. If a sufficient number of case studies are used then increases of channel slope, bankfull discharge, and therefore of stream power following channelization, delineate the general magnitude of potential post-construction adjustments. Figure 9.4 plots bankfull discharge and slope for sites along several straightened channels in Denmark

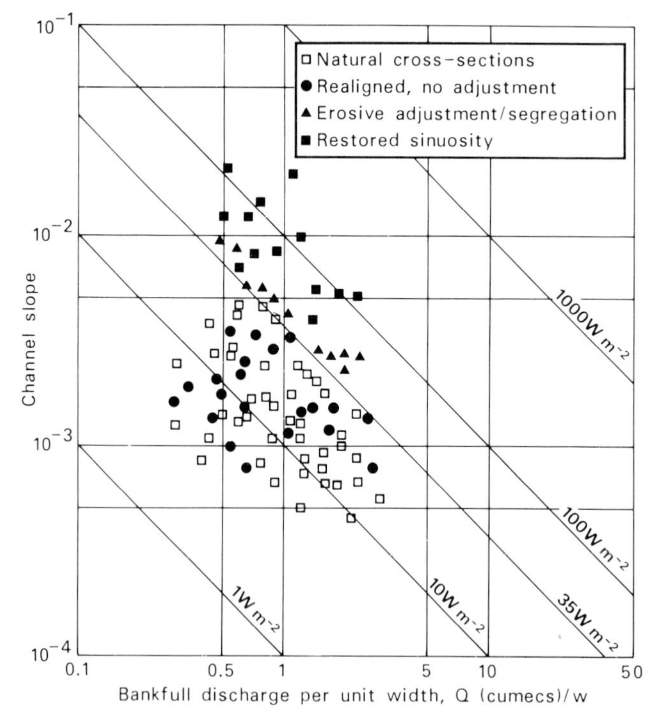


**Figure 9.4** Channel stability of Danish streams related to specific stream power. Each site is plotted as a function of bankfull discharge and channel slope, and lines of equal specific stream power have been superimposed (after Brookes, 1987b)

and this indicates how a threshold stream power of 35 W m$^{-2}$ discriminates fairly well between stable and unstable sites (Brookes, 1987b). Cross-sections for natural meandering streams lie below the 35 W m$^{-2}$ line and this reflects the feeble low energy nature of Danish streams and the fact that their channels are not therefore actively migrating. A second category of channels also falls below the 35 W m$^{-2}$ threshold and, although the slope of these channels was increased by artificial straightening, they have not undergone erosive adjustment. However, points above the threshold represent channels which have either eroded their bed and banks as a consequence of straightening, or have attempted to regain their former sinuosity. Channels stable prior to straightening are more likely to remain stable, but channels which are initially unstable will probably require extensive engineering and maintenance following construction. Relationships such as the one in Figure 9.4 can be employed to indicate which situations are prone to major adjustments and as the database (Figure 9.3b) is enlarged further, discriminant relationships can be produced.

In the absence of periodic remodification of channels or maintenance, morphological and biological recovery may occur naturally. For example, a straightened reach which regains sinuosity and develops a series of pools and riffles may become biologically productive. There is not yet a sufficient number of case studies available to make accurate predictions about recovery following channelization. However recovery appears to be a very slow process, and documented studies indicate time periods between ten and 100 years (Corning, 1975). After thirty years an enlarged reach of the Chariton River in Missouri had reverted from a uniform width and depth back to natural conditions with meanders and pools (Congdon, 1971). However, Arner (1975) found little recovery of fish productivity even 43 years after channelization of the Luxapalila River in Missouri.

## ALTERNATIVE PRACTICES

Clearly channelization has had significant direct and indirect impacts on rivers in several parts of the world and it is for this reason that some progress has been made towards developing alternative approaches. This new, less harsh philosophy has been emerging during the past fifteen years, particularly in North America, and involves less straightening and resectioning and is based instead on the emulation of the natural forms and processes of meandering channels. Table 9.3 summarizes the structures and actions employed in the planning, design, construction or maintenance of flood control channels that are intended to improve the instream environment. Although an understanding of fluvial processes and natural stream geometry is a vital ingredient of all successful environmental designs, geomorphologists have made a unique contribution to those alternatives which recreate elements of the natural channel morphology.

**Table 9.3** Environmental features used in flood control projects (based on Nunnally and Shields, 1985)

1. *Channel design and construction techniques*

 Selective clearing and snagging
 Low flow channels
 High flow channels
 Floodways and bypass channels
 Pools and riffles
 Meandering alignments
 Single bank modification

2. *Features for bed and bank stability and erosion control*

 Channel linings and bank protection measures
 Grade control structures
 Sediment and debris basins
 Sediment control during construction

3. *Features to improve instream habitat*

 Sills
 Deflectors
 Random rock structures
 Habitat cover devices
 Water level control structures
 Spawning beds
 Fishways

4. *Features for streamside aquatic and wetlands habitat*

 Maintenance of cutoff meanders
 Bendway design considerations
 Wetlands

Such alternatives have been conceived to produce a managed river which behaves as much like a natural one as possible, which looks as aesthetically pleasing as possible and which minimizes the morphological and ecological disruption which follows.

Research in West Germany advocated the preservation or construction of meanders and asymmetrical cross-sections to emulate the morphology of natural stream channels by directing flow at particular banks and inducing the formation of a series of pools and riffles (Seibert, 1960). Such techniques produce a diverse range of habitats, providing channels which are more biologically productive (Binder and Gröbmaier, 1978). Experiments undertaken on Gum Branch Creek near Charlotte in North Carolina showed how the channel cross-sections could be manipulated to influence stream processes and induce the development of point bars at desired locations (Keller, 1975; 1978). Bars developed where an asymmetrical cross-section was cut with a slope of 2:1 on one bank, and 3:1

on the opposite bank. This profile caused high flows to converge and scour the steeper bank, whilst the point bar would be deposited adjacent to the gentler bank. Symmetrical profiles with both banks at a slope of 2:1 were designed to diverge flows.

The use of varied cross-sections, rather than conventional uniform designs, is implicit in the concept of channel restoration or renovation, also developed in North Carolina (Keller and Hoffman, 1977; Nunnally, 1978). This is a means of restoring the flow efficiency of streams which have become choked with debris or have undergone severe bank erosion. Restoration is accomplished by the removal of debris jams, providing non-uniform channel cross-sections and gradients, whilst preserving meanders, leaving as many trees as possible along stream banks, and stabilizing banks with vegetation and riprap where necessary.

Where the engineering objectives of flood control and agricultural drainage are no longer required for a channel, then it is possible to recreate the former natural morphology. This has been achieved for a small straightened stream in southern Jutland, Denmark, where the sinuosity, cross-sectional dimensions, slope and substrate were reinstated in 1984 (Brookes, 1987c). The original course was recreated by reference to historical maps, excavation of depressions on the floodplain and observations made of adjacent natural stream channels.

## LEGISLATION AND POLICY CONSTRAINTS

A number of promising and effective schemes are now available to achieve the objectives of channelization without instigating the deleterious effects that have been criticized. However, the precise contribution that geomorphologists can make in specific countries is very much dependent on the legislation requirements and on the local policies which govern the management of watercourses. It is evident that the extent to which legislation has developed has varied significantly from one country to another (Table 9.4). In the United States there is a long history of legislation dating from the Rivers and Harbours Act of 1899 and subsequently including (Table 9.4) the Fish and Wildlife Coordination Act of 1958, the National Environmental Policy Act of 1969, and the Wild and Scenic Rivers Act of 1968. The latter two Acts involve requirements for managing the rivers of the nation particularly in relation to environmental impact statements. In New Zealand and Canada there have been other acts leading to different degrees of legislative control on watercourses.

In the case of drainage authorities in England and Wales the requirements are not as stringent as those for the United States. Under Section 22 of the Water Act of 1973 the provision that authorities should 'have regard' to the needs of conservation was of limited value and it was a significant advance when, in 1981, this was amended by the Wildlife and Countryside Act of 1981 to give authorities the responsibility to 'further' conservation. A number of guidelines have

**Table 9.4** Legislation and guidelines for channel management

| Country | Legislation and guidelines | Implications of legislation for channelization works |
|---|---|---|
| United States of America | Rivers and Harbours Act 1899 | Regulates the excavation and filling in any manner to alter or modify the course, location, condition or capacity of any navigable water |
| | Fish and Wildlife Coordination Act 1958 | Requires that wildlife be considered in the design and implementation of water resource development programme and the mitigation of development impacts |
| | National Environment Policy Act 1969 | Requires preparation and review of environmental impact statements for '*major Federal action significantly affecting the quality of the human environment*' |
| | Wild and Scenic Rivers Act 1968 | Designates selected rivers of the Nation, '*which with their immediate environments, possess outstandingly remarkable scenic, recreational, geologic, historic, cultural or other similar values, shall be preserved in free-flowing condition*' |
| Canada | Canada Water Act 1970 | Provides '*for the management of water resources of Canada including research and the planning and implementation of programmes relating to the conservation, development and utilization of water resources*' |
| | Clean Environment Act 1975 (New Brunswick) | Watercourse Alteration Regulation regulates '*any change made intentionally at, near or to a watercourse or waterflow in a watercourse, both temporary and permanent*' |
| New Zealand | Water and Soil Conservation Act 1967 | Requires that every board (Regional Water Board) have regard to recreational needs and the safeguarding of scenic and natural features, fisheries and wildlife |
| Denmark | Watercourse Act 1949, amended 63, 65, 69, 73 | Gives drainage priority over all other uses but, regulates dredging and the timing of weed control |
| | Environmental Protection Act 1974 | Requires bodies '*to safeguard environmental qualities essential to the . . . maintenance of the diversity of plants and animals*' |
| | Watercourse Act 1982 | Requires that works are to be planned and undertaken with regard to the stream quality (this includes the physical form of the stream, incuding the pools and riffles) |
| | | Supplement to the 1982 Act states the legal procedure for restoring watercourses |

*(continued)*

**Table 9.4** *(continued)*

| Country | Legislation and guidelines | Implications of legislation for channelization works |
|---|---|---|
| Great Britain | Water Act 1973 section 22 | Water Authorities must '*have regard to the desirability of preserving the natural beauty, of conserving the flora and fauna and geological or physiographical features of special interest*' |
| | Wildlife and Countryside Act 1981 | Extends the above requirements. Water Authorities and other bodies concerned with land drainage '*shall . . . so exercise their functions . . . as to further conservation and enhancement of natural beauty and the conservation of flora, fauna and geological and physiographical features of special interest*' |

subsequently been formulated for the management of rivers in England and Wales, most notably those produced by the Royal Society for the Protection of Birds and the Royal Society for Nature Conservation (Lewis and Williams, 1984). The imminent requirement for environmental impact statements, arising from European Community legislation, should lead to a greater significance being attached to these developments.

In some European countries legislation has dictated that consideration should be given to the morphology of the channel and the Danish Watercourse Act of 1982 states that river works are to be planned and undertaken with regard to the physical form, including pools and riffles (Brookes, 1987b). Furthermore the Watercourse supplement of 1983 has allowed for the restoration of watercourses using techniques which are more sympathetic to the natural environment.

## INTERDISCIPLINARY RIVER MANAGEMENT IN THE BASIN CONTEXT

The extent to which alternatives to channelization schemes need to be considered as part of public policy necessarily varies according to legislation currently in force. However, it is now realized that there are considerable benefits to be gained by adopting an interdisciplinary approach to river management and by focusing upon management in the context of the drainage basin as a whole.

Within the UK problems of flood control and land drainage have traditionally been solved by an engineering design (Figure 9.5). However, environmental and planning issues have made the implementation of river schemes a contentious and increasingly difficult operation in recent years. Thames Water has therefore recently developed a new interdisciplinary approach to river management which

1. TRADITIONAL ROUTE

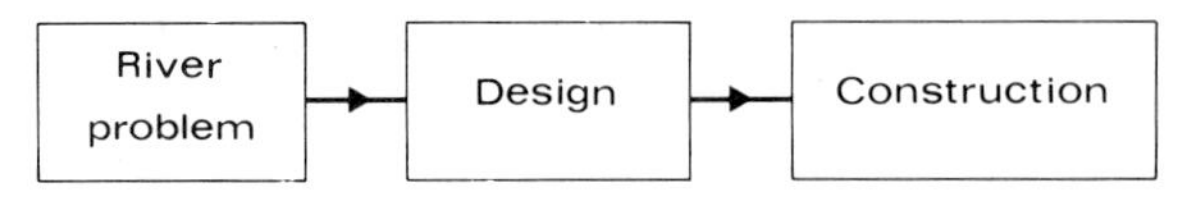


2. HOLISTIC APPROACH

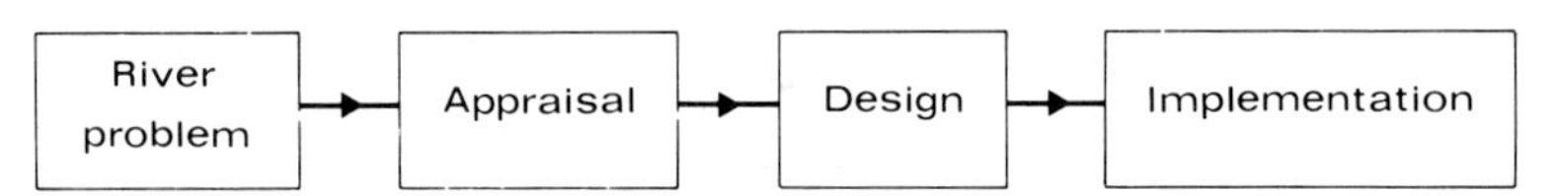


**Figure 9.5** Approaches to channel management

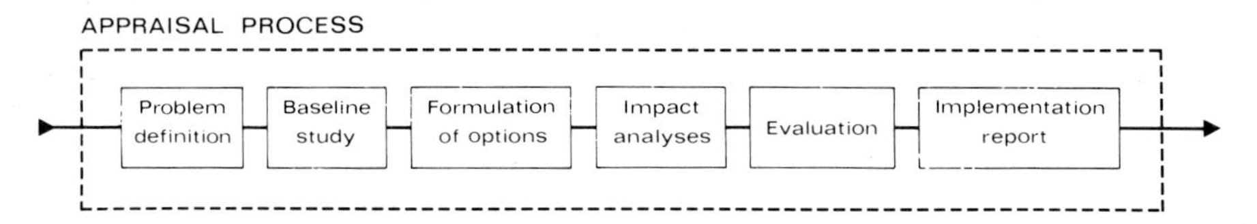


**Figure 9.6** Stages in the appraisal process

is unique to the UK Water Industry. Through this holistic approach, which is indicated by route 2 (Figure 9.5) potential schemes are now appraised prior to design in terms of environmental, political and economic factors, including impact assessments. It is within this framework that geomorphic approaches to river management are being developed by Thames Water. It is believed that detailed appraisal creates a better basis for the decision-making process and ensures a useful programme and structure for scheme development.

A four-stage appraisal process has been adopted (Figure 9.6) whereby the nature of the problem is defined, baseline studies are then undertaken which involve collection of all the necessary data; a range of alternative options, including the 'do-nothing' state, are then evaluated, before a final option is selected.

The Appraisal Team within Thames Water is producing an Appraisal Manual for the guidance of river managers, and this manual necessarily reflects interdisciplinary expertise. The first chapter is devoted to geomorphology and contains advice on geomorphic appraisal, including changes which occur at the scale of an individual reach and within the river basin. Attention is also given to post-project appraisal which aims to quantify the impacts of existing engineering schemes. The chapter has drawn heavily on North American and European experience and techniques in the absence of sufficient information specific to the UK. The lack of an adequate database has prompted a catchment study for the Thames Basin, utilizing the findings and expertise of a number of fluvial geomorphologists and one product will be a sensitivity map to indicate the types of engineering design suited to a particular channel type. Such approaches can then provide information to allow public policy decisions to be made in the most informed way.

In addition to employing an interdisciplinary approach it is increasingly necessary to envision channel management within the context of the drainage basin as a whole. Although the need for basin-based management of water resources has been accepted for some years, failure to consider the total fluvial system at the basin scale when planning a flood control project frequently leads to environmental problems. Both the hydraulic geometry and regime approaches to river channels are essentially empirical and consider streams as equilibrium systems. However, most channels are in a transient state as a result of human activities having an impact on channels either directly at the scale of a reach, or indirectly on catchment processes, altering the sediment and water discharges. It is important to understand the effects that changes in a basin will have on channelization projects and a number of functional relationships are available which indicate the nature and direction of channel adjustment (Lane, 1955; Schumm, 1969; 1977). In turn channel adjustments may trigger changes in the ecosystem.

An example is provided by the drainage basin of the Sungai Kelang, in which Kuala Lumpur is situated, in Malaysia. Channelization of the river has recently been completed through Kuala Lumpur using an impressive compound channel which is aesthetically imaginative. However, land development by deforestation upstream of Kuala Lumpur means that large sediment loads are being provided to move through this scheme from channels which are themselves eroding. Downstream of the channelized river the channel to the sea is experiencing active aggradation and channel capacities are decreasing (Gregory, 1987, see also Douglas, this volume, Chapter 5). This example emphasizes the need to adopt a drainage basin approach to river channel management and this is further underlined by the way in which changes of channel capacity can affect the capability of a channel to accommodate discharges. Thus downstream from Kuala Lumpur increased discharges, increased partly by the channelization scheme, led to channel erosion and to larger channel capacities. Subsequently, however, the increased sediment production from cleared areas, together with urban sediment provision has led to aggradation which gives lower channel capacities less able to accommodate the greater flood discharges. It is therefore essential to adopt a basin approach which embraces consideration of the multifarious dimensions of change.

Some progress towards the classification of drainage basins was made by Palmer (1976) for Washington and Oregon, where steep mountain streams descend into populated lowlands and empty into tidewaters or major rivers. The rivers were classified into four geo-hydraulic zones, namely the boulder, floodway, pastoral and estuarine zones. These four categories were used to define the long-term (thousands of years) geomorphic processes and short-term channel hydraulic processes, and the land uses that can be supported in each zone. Management guidelines were developed for the entire system, with the intention of sustaining the equilibrium of natural processes. For example, the floodway zone develops a valley floor by the river channel shifting across the floodplain. In this zone the supply of sand and gravel is balanced by the ability of the stream to transport sediment and a moderately steep slope provides the energy for this transport. Channel features having a high value for wildlife habitat and scenic and recreational use, include pools, riffles and point bars. Since this is a relatively high energy zone with low bank resistance, attempts to straighten and narrow the channel will intensify the velocities and the equilibrium that the channel acquired over thousands of years will be disrupted. It is recommended (Palmer, 1976) that the best way to protect the river zone is to adopt a natural streamway corridor, where boundaries are defined and strict land use controls to prevent development are enforced.

Information from channelization schemes in many areas of the world (Figure 9.3(a),(b)) together with general relationships (e.g. Figure 9.4) should now be developed to provide a more precise and process-based basin approach to channel management. It is necessary to refine such basin approaches and extend them to other countries, including the UK. A basin approach to channel management and design will assist more efficient and more economic management, provide a reference database reflecting drainage basin characteristics and hence should afford a frame of reference for developers and for the wider community who have to accept river channel management methods.

## CONCLUSION

Research on channelization by geomorphologists has taken place to a considerable degree over the last two decades. In that time we have progressed from the recognition of the impact of channelization schemes both within and adjacent to the scheme, to the formulation of alternative measures of channelization. Public policy has been pertinent to specific reaches in the management of channelization until recent years. However, subsequently it has been realized that the policy decisions made in relation to particular reaches have implications for other areas. These other implications which cover morphological and ecological effects of channelization also extend to the aesthetic quality of the river environment. It is for these reasons therefore that the present chapter has argued for a more drainage basin based approach to river management embracing river channelization.

Contributions made by geomorphologists to public policy are perhaps most notable in the United States, where legislation requiring environmental impact assessment has been established since 1969. This influence is represented by published material which is used by government agencies responsible for river management, by individuals working in industry, and through consultancy. Although legislative requirements are less stringent in the UK it is likely that the contribution of geomorphology will increase following the introduction of EIA requirements in 1988.

One ingredient that unites fluvial geomorphology and river engineering is the understanding of how a given action by man will produce feedback changes that can alter other components of the natural system. The behaviour of river channels can be deduced by observing responses of stream systems that have already been altered. The success of alternative practices to conventional engineering designs relies on an understanding of fluvial processes and forms. Clearly there is a need to formulate designs and channel management practices which consider the whole basin rather than isolated channel reaches, and this is one of the major bridges that needs to be built between geomorphology and engineering as a basis for guiding public policy decisions.

### Acknowledgement

The assistance given by Mr P. A. Burkhard in the collection of data for Figure 9.3 and by Mr C. T. Hill in the compilation of some data in Table 9.4 is gratefully acknowledged. The views expressed are those of the authors and not necessarily those of Thames Water.

## REFERENCES

Arner, D. H. (1975). Report on the effects of channelization modification on the Luxapalila River, *Symposium on Stream Channel Modification*, Harrisonburg, Virginia.

Bayless, J. and W. B. Smith (1967). The effects of channelization upon the fish population of lotic waters in eastern North Carolina, *Proc. Annual Conference S.E. Assoc. Game and Fish Comm.*, **18**, 230–38.

Binder, W. and W. Gröbmaier (1978). Bach- und flusslaufe-ihre gestalt und pflete, *Garten U. Landschaft.*, **1**, 78, 25–30.

Bird, J. F. (1980). Geomorphological implications of flood control measures: Lang Lang River, Victoria, *Australian Geographical Studies*, **18**, 169–83.

Brice, J. C. (1981). *Stability of Relocated Stream Channels*, Report prepared for the Federal Highway Administration, No. FHWA-RD-80-158.

Brookes, A. (1984). *Recommendations Bearing on the Sinuosity of Danish Stream Channels*, National Agency of Environmental Protection, Silkeborg, Denmark, Technical Report No. 6.

Brookes, A. (1985a). River channelization: traditional engineering practices, physical effects and alternative practices, *Progress in Physical Geography*, **9**, 44–73.

Brookes, A. (1985b). Downstream morphological consequences of river channelization in England and Wales, *Geographical Journal*, **151**, 57–62.

Brookes, A. (1987a). River channel adjustments downstream from channelization works in England and Wales, *Earth Surface Processes and Landforms*, **12**, 337–51.

Brookes, A. (1987b). The distribution and management of channelized streams in Denmark, *Regulated Rivers*, **1**(1), 3–16.

Brookes, A. (1987c). 'Restoring the sinuosity of artificially straightened stream channels, *Environ. Geol. Water Sci.*, **10**, 33–41.

Brookes, A., K. J. Gregory and F. H. Dawson (1983). An assessment of river channelization in England and Wales, *The Science of the Total Environment*, **27**, 97–112.

Committee on Government Operations (1973). *Stream Channelization: what Federally Financed Draglines and Bulldozers do to our Nation's Streams*. Fifth report together with additional views, *House Report* 93–530, Washington DC, 139pp.

Congdon, J. C. (1971). Fish populations of channelized and unchannelized sections of the Chariton River, Missouri. In E. Schneberger and J. E. Funk (eds) *Stream Channelization—a Symposium*, pp. 52–62, North Central Division American Fish Society.

Corning, R. V. (1975). Channelization: shortcut to nowhere, *Virginia Wildlife*, **6**, 8.

Daniels, R. B. (1960). Entrenchment of the Willow Creek Drainage Ditch, Harrison County, Iowa, *American Journal of Science*, **258**, 161–76.

Dodge, W. E., E. E. Possardt, R. J. Reed and W. P. MacConnell, *Channelization Assessment, White River, Vermont: Remote Sensing, Benthos and Wildlife*, Report No. FWS-OBS-76/7, Office of Biological Services, U.S. Fish and Wildlife Service, US Department of the Interior, Washington DC., 74pp.

Emerson, J. W. (1971). Channelization: a case study, *Science*, **173**, 325–6.

Graham, R. (1975). Physical and biological effects of alterations on Montana's trout streams, Paper presented at *Symposium on Stream Channel Modifications*, Harrisonburg, Virginia.

Gregory, K. J. (ed.) (1977). *River Channel Changes*, John Wiley & Sons, Chichester. 450pp.

Gregory, K. J. (1979). Hydrogeomorphology: how applied should we become? *Progress in Physical Geography*, **3**, 84–100.

Gregory, K. J. (1987). Hydrogeomorphology of river channel changes, *Ilmu Alam* in press.

Griggs, G. B. and L. Paris (1982). The failure of flood control on the San Lorenzo River, California, *Environmental Management*, **6**, 407–419.

Griswold, B. L. (1978). *Some Effects of Stream Channelization on Fish Populations, Macroinvertebrates, and Fishing in Ohio and Indiana*, Report No. FWS-OBS-77/46, Office of Biological Services, Fish and Wildlife Service, US Department of the Interior, Washington DC.

Groen, C. L. and J. C. Schmulbach (1978). The sport fishery of the unchannelized Middle Missouri River, *Transactions American Fish Society*, **107**, 412–18.

Hansen, D. R. and R. J. Muncy (1971). *Effects of Stream Channelization on Fish and Bottom Fauna in the Little Sioux River, Iowa*, Iowa State Water Resources Institute, 38.

Headrick, M. R. (1976). *Effects of Stream Channelization on Fish Populations in the Buena Vista Marsh, Portage County, Wisconsin*. M. S. Thesis, University of Wisconsin.

Huggins, D. G. and R. E. Moss (1974). Fish population structure in altered and unaltered areas of a small Kansas stream, *Transactions of the Kansas Academy of Science*, **77**, 18–30.

Keller, E. A. (1975). Channelization: a search for a better way, *Geology*, **3**, 246–8.

Keller, E. A. (1976). Channelization: environmental, geomorphic and engineering aspects. In D. R. Coates (ed.) *Geomorphology and Engineering*, pp. 115–140, George Allen & Unwin, London.

Keller, E. A. (1978). Pools, riffles and channelization, *Environmental Geology*, **2**, 119–27.
Keller, E. A. and E. K. Hoffman (1977). Urban streams: sensual blight or amenity, *Journal of Soil and Water Conservation*, **32**, 237–42.
Lane, E. W. (1947). The effect of cutting off bends in rivers, *Proceedings of the Third Hydraulic Conference*, Bulletin 31, Iowa.
Lane, E. W. (1955). The importance of fluvial morphology in hydraulic engineering, *American Society of Civil Engineers*, **81**, 741, 181–97.
Leopold, L. B. (1977). A reverence for rivers, *Geology*, **5**, 429–30.
Lewis, G. and G. Williams (1984). *Rivers and Wildlife Handbook: a Guide to Practices which Further the Conservation of Wildlife on Rivers*, Royal Society for the Protection of Birds, Sandy, Bedfordshire, England, 296pp.
Morris, L. A., R. M. Langemeirs, T. R. Russel and A. Witt (1968). Effects of main stem impoundments and channelization upon the limnology of the Missouri River, Nebraska, *Transactions American Fish Society*, **97** (4), 380–88.
Moyle, P. B. (1976). Some effects of channelization on the fishes and invertebrates of Rush Creek, Modoc County, California, *California Fish and Game*, **62**(3), 179–86.
Newson, M. (1986). River basin engineering—fluvial geomorphology, *Journal of the Inst., of Water Engrs and Sci.*, **40**, 307–24.
Nixon, M. (1966). Flood regulation and river training. In R. B. Thorn (ed.) *River Engineering and Water Conservation Works*, Butterworths, London.
Nunnally, N. R. (1978). Stream renovation: an alternative to channelization, *Environmental Management*, **2**, 403–11.
Nunnally, N. R. and F. D. Shields (1985). *Incorporation of Environmental Features in Flood Control Channel Projects*, Technical Report No. E-85-3. US Army Engineer Waterways Experiment Station, Vicksburg, Mississippi.
Palmer, L. (1976). River management criteria for Oregon and Washington. In D. R. Coates (ed.) *Geomorphology and Engineering*, pp. 329–46, George Allen & Unwin, London.
Parker, G. and D. Andres (1976). Detrimental effects of river channelization, *American Society of Civil Engineers, Rivers '76*, 1248–66.
Ruhe, R. V. (1970). Stream regimen and man's manipulation. In D. R. Coates (ed.) *Environmental Geomorphology*, Binghamton, State University of New York, Publication in Geomorphology. pp. 9–23.
Schmal, R. N. and Sanders, D. F. (1978). *Effects of Stream Channelization on Aquatic Macro-invertebrates, Buena Vista Marsh, Portage Country, Wisconsin*, Fish and Wildlife Service, United States Department of the Interior.
Schumm, S. A. (1969). River metamorphosis, *Proceedings of the American Society of Civil Engineering, Journal of Hydraulics Division*, **95**, 255–73.
Schumm, S. A. (1977). *The Fluvial System*, Wiley, New York.
Schumm, S. A., M. D. Harvey and C. C. Watson (1984). *Incised Channels: Morphology Dynamics and Control*, Water Resources Publications, Littleton, Colorado.
Seibert, P. (1960). Naturnahe Querprofilgestaltung bei Anbau von Wasserlaufen, *Natur U. Landschaft,* **35**, 12–13.
Szilagy, J. (1932). Flood control on the Tisza River, *Military Engineer*, **24**, 632.
Winkley, B. R. (1972). River regulation with the aid of nature, *International Commission on Irrigation and Drainage*, Eighth Congress, pp. 433–72.
Yearke, L. W. (1971). River erosion due to channel relocation, *Civil Engineering*, **41**, 39–40.

Geomorphology in Environmental Planning
Edited by J. M. Hooke



# 10 Urban River Pollution in the UK: the WRc River Basin Management Programme

**BOB CRABTREE**
*WRc Engineering, Swindon, Wiltshire*

## INTRODUCTION: THE INTERACTION BETWEEN URBAN DRAINAGE SYSTEMS AND RIVERS

### The Background

The results of the 1985 River Quality Survey of England and Wales showed that some 11 per cent of all freshwater rivers and canals fell into the quality categories of 'poor' or 'grossly polluted' (HMSO, 1986); a total length of 4210 kilometres. For the first time since the commencement of regular surveys in 1958 the proportion falling into these two categories increased. A recent survey of perceived river quality problems and outstanding modelling needs (Crabtree, 1986) placed urban runoff pollution and non-point pollution in rural areas as top priorities for the Regional Water Authorities and other responsible agencies. For example, North West Water have identified the need for expenditure of £3.7 billion over 25 years (North West Water, 1983) to upgrade their surface waters to acceptable quality levels. Other regions have similar problems in nature if not in extent.

Examination of the changes in water quality between 1980 and 1985 (HMSO, 1986) and the general spatial pattern of water quality shows two clear trends in relation to the identified priority sources of pollution. First the deterioration of previously good quality rural streams particularly in the west of the country and second the persistence of the 'poor' and 'grossly polluted' watercourses, concentrated around the older industrial conurbations particularly in Lancashire, Yorkshire and the Midlands.

The aim of this chapter is to discuss the significance and causes of pollution in the second of these, the urban watercourses. In particular, the contribution of urban drainage systems to river pollution is presented and collaborative research at the WRc into urban river pollution control, within the River Basin Management Programme, is described. Upon completion, the products of this programme (Clifforde, Saul and Tyson, 1986) will enable the water industry to manage the quality of urban rivers and carry out pollution control measures in a more objective and cost-effective manner than has been possible in the past. This will only be possible by the adoption of a policy of integrated river basin management which takes due consideration of the hydrological, geomorphological and biochemical processes which operate in natural and urban catchments.

## Development of Urban Drainage Systems

Figure 10.1 is a schematic plan of the typical drainage system of many towns in the United Kingdom. This plan illustrates the general features of an urban drainage system and the changes in design practice that have occurred over time.

The first widespread development of piped or covered drainage systems took place following the growth of industrial towns in Victorian times. They were built to contain the original open ditch and channel systems which had become unacceptable with increased population density and the increased use of

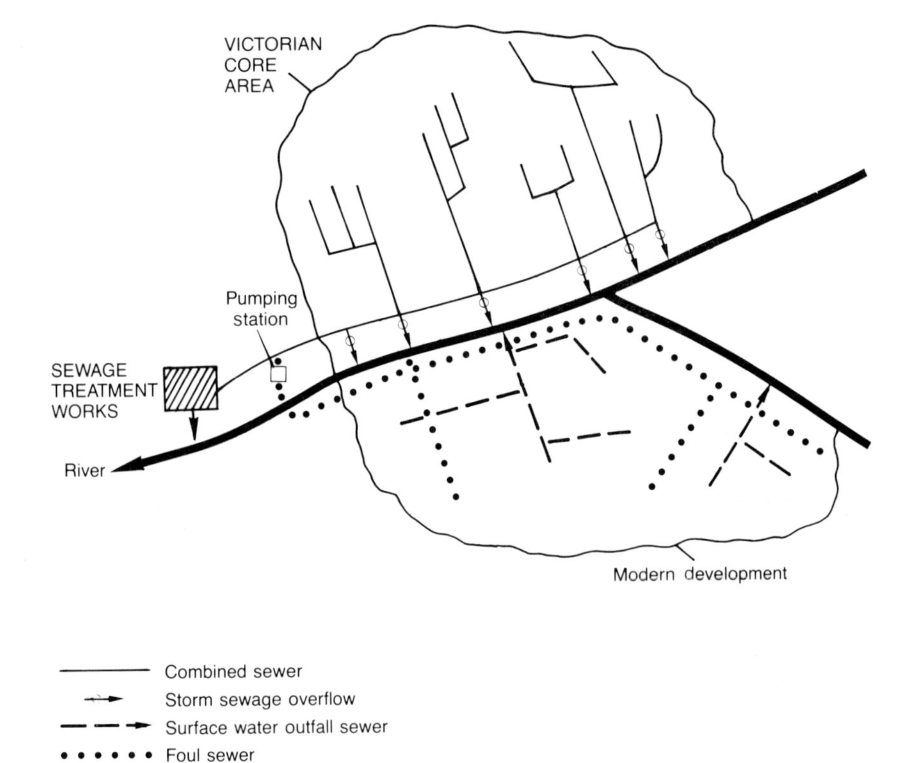


**Figure 10.1** A typical urban drainage system in the UK. Reproduced by permission of the Water Research Centre

water-based domestic waste removal. Thus the sewers and culverted watercourses in the Victorian core areas of towns carried both foul sewage from industrial and domestic sources and also surface runoff in times of rainfall.

The development of sewage treatment works at the ends of such 'combined' sewerage systems followed on from this first development phase when the receiving watercourses become unacceptably polluted to the extent of becoming a general hazard to public health. Financial and technical constraints limited the size of sewers and sewage treatment facilities and this in turn restricted the amount of sewage that could be carried by the sewerage system and treated by the works. To prevent flooding from these combined sewers, overflows were introduced to pass excess flows directly to a watercourse. It was considered that such overflows would only operate in times of rainfall when the foul sewage was considered to be sufficiently diluted with clean surface water.

As urban areas expanded efforts were made to reduce the volume of sewage within the original combined sewerage system by reducing the input of surface water. This was to reduce the amount of water being carried to treatment works and the amount of pollution spilled to rivers by overflows. The partitioning of sources led to the 'separate' systems of drainage typical of recent and modern development. In such a separate system all the foul sewage is passed to treatment via a foul sewer which should have no overflows. The surface drainage is carried

**Figure 10.2** An unsatisfactory SSO in operation. Reproduced by permission of North West Water

via a surface water sewer to a watercourse. In practice, it is common for such separate systems to become accidentally cross-connected.

There are three possible sources of pollution from a drainage system. These are the discharge from the sewerage treatment works itself; discharges from surface water drains and discharges from storm sewage overflows (SSO) on combined sewers. Figure 10.2 illustrates an example of a unsatisfactory SSO discharge.

## The Basis for the WRc River Quality Management Programme

While the three main sources of urban pollution have been identified, the nature of their behaviour also needs to be considered. Many urban rivers are polluted by the continuous discharges of effluent from sewage treatment works and industrial plant. These continuous discharges can be controlled by means of discharge consents, within the legislative framework of the Control of Pollution Act 1974, Part II (COPA-II), (Matthews, 1987). Adequate methodologies exist for consent setting for continuous discharges (Crabtree and Cluckie, 1987). These involve the use of statistical techniques and deterministic river quality models (Crabtree *et al.*, 1987). There are at present no similar quantitative methods for controlling storm sewage and surface water outfall discharges. These discharges are intermittent in nature and the transient pollution they may cause can have both short-term, acute and long-term, chronic effects on the downstream river quality. To date these discharges are consented, within the

**Table 10.1** Sources and control of urban river pollution

| | |
|---|---|
| (A) *Urban drainage systems* | |
| 1. Combined sewers: | —domestic sewage<br>—industrial sewage<br>—surface runoff |
| 2. Separate sewerage system:<br>Foul sewer | —domestic sewage<br>—industrial sewage |
| Storm sewer | —surface runoff |
| (B) *Control of urban river pollution* | |
| 1. Continuous discharges from: | —Sewage treatment works<br>—Industrial effluents |
| Controlled by COPA-II with quantitative consent to enable downstream river quality objective to be achieved. | |
| 2. Intermittent discharges from: | —Combined sewer overflows (SSO)<br>—Surface water sewers |
| Only controlled by qualitative consents to prevent nuisance. | |

**Table 10.2** Mean pollutant concentrations of foul sewage, stormwater and storm sewage in the United Kingdom. Reproduced by permission of The Water Research Centre

| | Foul sewer | Surface water sewer | Combined sewer |
|---|---|---|---|
| Biological oxygen demand | 400 | 15 | 100 |
| Chemical oxygen demand | 750 | 100 | 380 |
| Ammoniacal nitrogen | 30 | 2 | 8 |
| Suspended solids | 300 | 190 | 400 |

Concentrations expressed in milligrams per litre.

framework of COPA-II, in a qualitative sense. For example, the consent may be that the overflow or outfall should only discharge during times of rainfall. Table 10.1 summarizes the different nature and means of control for the various major sources of urban river pollution. Table 10.2 indicates the mean pollutant concentrations associated with surface water outfalls and SSO discharges.

Generally increased public awareness of environmental issues and specifically, the possibility of privatization of the water industry in the near future have highlighted the need for more effective techniques, both for regulatory purposes and for planning and management activities. At present these techniques do not exist for controlling intermittent discharges. It is the purpose of the WRc River Basin Management (RBM) Programme, supported by the water industry, to develop these techniques. The present programme is principally directed towards providing the methodology for the water industry to deal with transient river pollution caused by spillage from combined sewers via SSOs. In the longer term, the aim is to produce a methodology for integrated river catchment quality planning and management which will provide a framework for controlling pollution from both continuous and intermittent, discrete and diffuse sources.

## SEWERAGE REHABILITATION: CONCEPT AND APPROACH

### The Scale of Pollution from Storm Sewage Overflows (SSOs)

Combined sewer systems, when effectively designed and operated, need not discharge storm sewage with such frequency or in such large volumes as to cause unacceptable pollution of the receiving water. However, the age of most UK systems, the piecemeal way in which they have been extended and the limitations of currently available investigative tools for planning, maintenance and operation, mean that excessive pollution through poorly controlled spillage of storm sewage does occur in many cases.

SSO pollution is infrequent and transient but in many cases may be of sufficient magnitude, in terms of concentration and load to be the critical limiting factor on river ecology (Bowden and Solbe, 1986; Hill, 1984). SSOs are seen by Water Authorities to be a major contributing factor to the unsatisfactory quality of many urban watercourses. North West Water, in a recent review (North West Water, 1983) believed that some 250 km of their rivers (5 per cent of the total of 5323 km) are unable to achieve their quality objectives because of the effects of SSOs.

On a national scale, the Ministry of Housing and Local Government Technical Committee on Storm Overflows and the Disposal of Storm Sewage concluded in their Final Report (HMSO, 1970) that there were between 10 000 and 12 000 SSOs in England and Wales and that by a range of fairly subjective criteria, almost 40 per cent might be regarded as being unsatisfactory. A similar survey carried out by the Scottish Development Department (HMSO, 1977) identified more than 2000 SSOs in Scotland. Of these some 20 per cent were positively considered to be unsatisfactory and a further 20 per cent were so vaguely understood as to be unclassified. It is considered that these figures are probably underestimates of the true picture.

At present there are very little data to quantify the degree of pollution from SSOs in the UK. However, a recent study in New York (Roswell and Gaffoglio, 1986) has produced some quantitative data which are relatively comparable to the UK for the partitioning of pollutant sources. This major study of the effects of SSOs in New York found that for a typical year, during storm periods, only 10 per cent of total sewer flows were passed to treatment facilities, discharges from SSOs direct to rivers accounted for 57 per cent of the flow and the remaining 33 per cent was discharged to rivers from storm sewers. SSOs contributed 33 per cent of the total annual pollution load discharged to rivers.

## Present Practice in the Design and Operation of SSO

The fundamental design criterion for SSOs is the overflow setting. This is the level of flow within the sewer at which storm sewage begins to pass over the overflow to a watercourse. For a given storm event this setting will influence when spill will start to occur and the total volume of spill.

Traditionally SSO setting has been based principally upon sewerage criteria by considering the desired or practicable maximum carry-on flow in the sewer. Flows in excess of this capacity have been discharged to watercourses without consideration of the assimilative capacity or desired quality. This approach led to the establishment of fixed criteria, such as multiples of the dry weather sewer flow. For example 6 × DWF has been commonly used as the desired carry-on flow to treatment. This approach makes no allowance for variations in the strength of crude sewage between sites, nor the dilution ratio of the receiving watercourse. The 'Formula A' approach (HMSO, 1970) is an extension of this concept.

Formula A is expressed as:

$$\text{required carry-on flow} = \text{DWF} + 1350P + 2E \text{ litres per day}$$

where DWF is the dry weather flow in sewer (litres per day), $P$ is the population and $E$ is the industrial discharges (litres per day).

In the UK, this is probably the most extensively used criterion for the setting of SSOs. It would seem to be reasonable to assume that river pollution would be significantly reduced if all SSOs operated in accordance with this setting, although many 'blackspots' would still exist.

Recently, storage volumes have frequently been incorporated into SSOs. These improve performance in two ways. First, for a given spill event, the volume of spill is reduced and second the frequency of spill events is reduced. Therefore, as illustrated in Figure 10.3, storage is a relatively simple and cost effective way of reducing SSO pollution by keeping more flow within the sewer.

Monitoring of changes in the quality of sewage during storm events has indicated that there can be marked temporal variations in the strength of sewage (Stotz and Krauth, 1984; Thornton and Saul, 1986). Many combined sewer systems have been shown to exhibit a marked 'first foul flush' of pollutants at

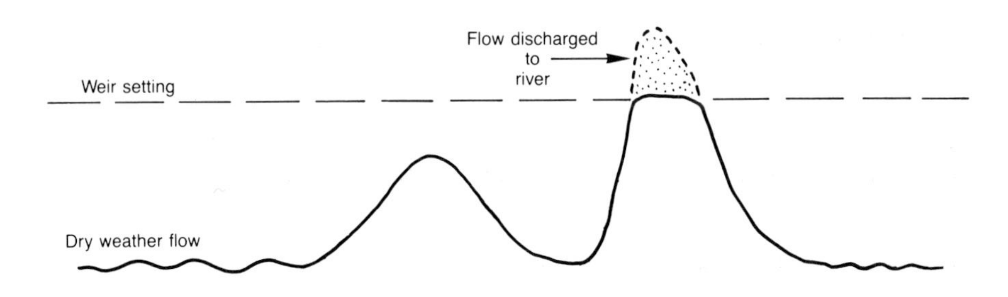


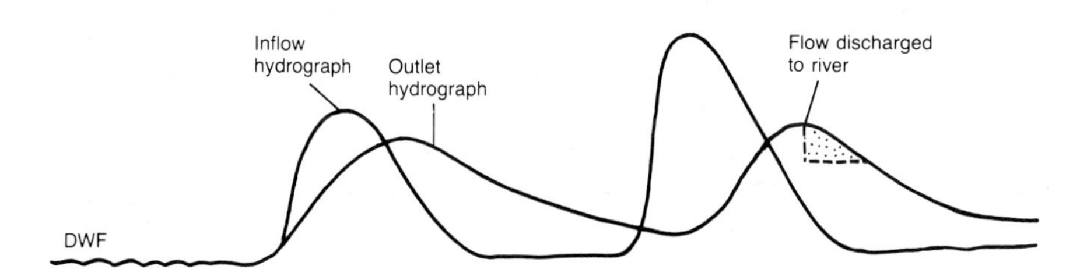


**Figure 10.3** The effect of storage volume on SSO operation. Reproduced by permission of the Water Research Centre

the onset of storm flow. Incorporating storage into SSOs can be an effective means of retaining this 'first foul flush'. Guidelines have been produced for the design of such storage overflows, which attempt to store the first flush and spill the subsequent, cleaner storm water (Ackers, Harrison and Brewer, 1968).

## Sewerage Rehabilitation for Pollution Control

The first edition of the WRc/WAA *Sewerage Rehabilitation Manual* (WRc/WAA, 1983) was mainly concerned with the hydraulic and structural upgrading of sewerage systems. SRM-II, the second edition of the *Sewerage Rehabilitation Manual* (WRc/WAA, 1986) provides an interim methodology for the control of pollution from SSOs, pending the development of more robust techniques by the RBM programme. The philosophical approach behind the interim and eventual methodology is that, while existing sewers can only carry a finite maximum rate of flow, a river can only accept a certain maximum amount of storm sewage if its quality objective is to be achieved. Hence surplus storm sewage must be stored within the sewerage system or diverted elsewhere. It is believed that efficient hydraulic design of SSOs will significantly reduce the problems of aesthetic pollution caused by the spillage of large objectionable solids and their subsequent stranding within watercourses. There are three main recommendations for the rehabilitation of SSOs. These are (WRc/WAA, 1986):

**Figure 10.4** A large modern overflow with storage. Reproduced by permission of the Water Research Centre

(1) The selection of an appropriate location for a SSO on an overloaded sewerage system and the capacity of the receiving watercourse to accept the discharge.
(2) The setting of the SSO to fully utilize downstream hydraulic capacity within the sewer and to minimize the spillage.
(3) The efficient design of each SSO to ensure hydraulic control and the maximum separation of gross polluting matter.

Individual SSOs should not be considered in isolation but in combination for each sewerage system. This can only be carried out by using a sophisticated hydraulic model of the sewer system, such as WASSP–SIM (NWC/DoE, 1981). Cost effective rehabilitation may be achieved by combining groups of small SSOs into a single large modern structure. This may facilitate the provision of storage to meet pollution performance criteria. Figure 10.4 illustrates a large modern SSO with storage.

## PRODUCTS OF THE WRc RBM PROGRAMME

For effective design of SSO and associated storage requirements to control river pollution and for scarce capital resources to be employed to the best effect, a rigorous working methodology is required which recognizes the limitations of the sewerage system and also those of the receiving water. To this end four major products are necessary to allow the objective planning of discharges from both SSOs and the less polluting but still significant surface water systems. These are:

(1) appropriate rainfall inputs to sewer flow simulation models;
(2) a sewer flow quality simulation model;
(3) a river impact model;
(4) a river quality classification system which recognizes the significance of intermittent pollution events.

The following section outlines the progress and ongoing developments in each of these four areas within the RMB programme.

### Rainfall Time Series for Sewer System Simulation Modelling

The objective of this work was to produce an annual series of fine resolution, historic rainfall data for use with the WASSP–SIM sewer hydraulic analysis model (NWC/DoE, 1981). The traditional approach to the use of sewer flow models has embodied the concept of design storms of standard profile and of a return period equal to the frequency with which exceedence of hydraulic capacity could be tolerated. While such design rainfalls represent extreme events they give little indication of the day-to-day performance of the sewerage system. To assess the performance and polluting impact of SSOs, the magnitude, frequency and duration of spills are required over a long time period, such as a typical year. This can be carried out using a flow simulation model, for example WASSP–SIM, by inputting a temporal sequence of rainfall events. Such a sequence, termed Time Series Rainfall (TSR) has been produced (Henderson, 1986). Two comparatively crude series, each representing a typical year, have been statistically synthesized for the east and west of the UK. These series were produced from 125 years of minute interval rainfall data from five sites. These series, in effect one year design rainfall sequences, are considered to be acceptable tools for the investigation of SSO performance and now incorporate techniques for adapting the regional TSR to local circumstances. Each series is available in three forms:

(1) As a chronological series of all significant runoff generating events in a typical year.
(2) The same events ranked in order of descending severity.
(3) A pair of similarly ranked series for summer and winter periods.

While the present TSRs are suitable for current usage, work is in hand to develop stochastic models to extend the series beyond one year and to improve the computational and regionalization aspects.

### Sewer Flow Quality Simulation Model

The proposed model is to be capable of simulating the build up, removal and transport of specified pollutants both above ground and within the sewer. The model will be applicable to both surface water and combined systems. The model will simulate both the total pollutant load passing through the sewer system and also the quality of the storm sewage at any location within a sewer and at any time during a storm event or sequence of events. The main output will be SSO discharge pollutographs. These will form the basic input to the receiving watercourse impact model.

Existing sewer flow quality models which aim to simulate time varying pollutant behaviour, for example SWMM (Huber, 1984), require extensive field verification. Simpler planning models, such as SAMBA (Johansen, 1983), produce event-based total load transport estimates. These are unsuitable for UK application, where the short-term impact of oxygen depletion and toxic substances is considered to be most important for the quality of the receiving water. This has led to the proposed development of a deterministic type model to simulate the processes occuring within the sewerage system. Eventually a statistical element will be introduced to obtain an estimate of confidence in the model results.

Such a model has two major subcomponents, based on the surface and subsurface drainage systems. The surface sub-system comprises three elements: sediment accumulation, sediment removal and pollutant-sediment associations. Sediment accumulation upon catchment surfaces has previously been represented by accumulation rates or by a continuous mass-balance approach. However, recent studies in the UK suggest that loads are a function of runoff transport capacity (Mance, 1982; Ellis, 1986). Consequently the proposed model will assume an unlimited sediment store for washoff. In addition, gulley pot sediment build up will be incorporated separately (Pratt, Elliott and Fulcher, 1986). Sediment transport will be simulated by a modified form of an existing sediment transport model (Price and Mance, 1978). Pollutant/sediment associations are used to generate pollutant levels in runoff, as related to the transport of different sediment size fractions, by the use of contaminant factors.

Within the subsurface drainage system, sediments from surface and foul sewage sources become mixed. Sediments are also available from in-pipe dry weather flow deposits. A scour and deposition sediment transport model is used to represent sediment behaviour within this system (Sonnen, 1977). Fundamental research is underway to investigate the behaviour of sediment within sewers. Until the results of this work become available the model will use an adaption of the Ackers–White sediment transport model (Ackers, 1984). Where necessary, contaminant degradation within the system will also be modelled. It is envisaged that the model must be capable of producing acceptable simulations of time-varying pollutant loadings without extensive site specific calibration. It is hoped to achieve model verification prior to application by carrying out a limited sampling and monitoring exercise to assess the local foul and storm sewage strengths.

## River Impact Model

The use of TSR and the development of the sewer flow quality simulation model will provide the necessary inputs to a river quality impact model. This model will predict the impact of SSO discharges on receiving waters for use in planning, designing and assessing the performance of sewerage rehabilitation schemes involving river pollution abatement. Initially the scope of the river impact model will be restricted to modelling the river in the immediate vicinity of an SSO, for the duration of spill events. Future enhancements will incorporate multiple overflows, downstream quality changes and long-term chronic impacts.

The immediate requirement is for a planning model to assess the immediate impact of SSO discharges in terms of short-term oxygen depletion and the concentrations of acutely toxic substances, such as ammonia and hydrogen sulphide. To perform this function, a dynamic deterministic model is required. The model needs to be able to make predictions at a timescale of minutes. It is considered that this timescale is significant in terms of the response of the

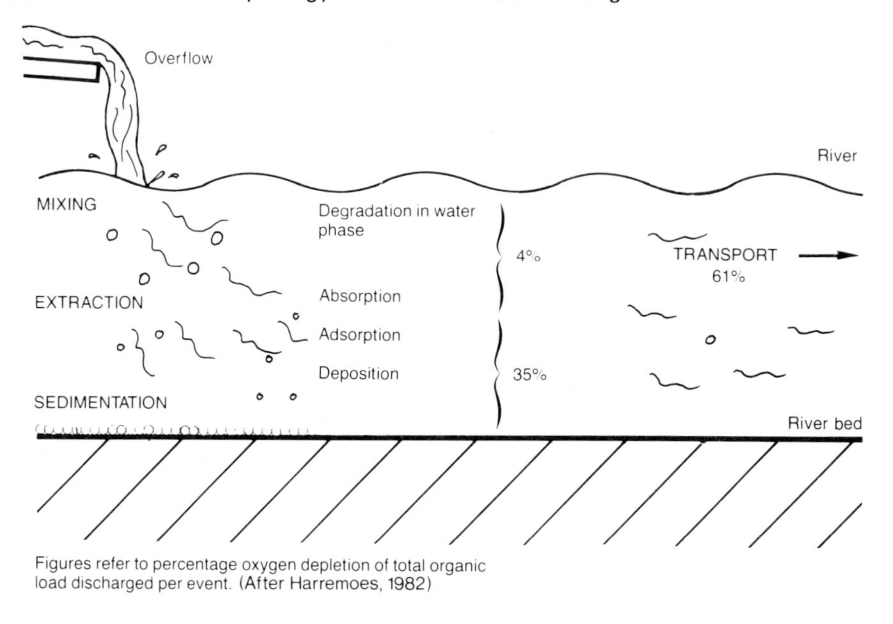


**Figure 10.5** Schematic removal processes for organic pollutants discharged to a river in the vicinity of an overflow (after Harremoes, 1982). Reproduced by permission of the Water Research Centre

river biota to high but transient levels of pollution. Outside this timescale, the frequency effects of transient pollution impacts must also be assessed. Recent detailed process studies in Denmark have investigated the impact of SSO discharges to small streams (Harremoes, 1982; Hvitved-Jacobsen, 1982). These studies identified three removal mechanisms for organic matter discharged from SSOs. These are illustrated in Figure 10.5. Of the total organic load discharged, only 39 per cent of the potential oxygen depletion occurred within the vicinity of the overflow. This could be further subdivided into 4 per cent exerted as an immediate oxygen demand in the stream due to mixing and adsorption and a further 35 per cent delayed demand due to absorption and sedimentation. The long-term effects of chronic oxygen depletion and nutrient enrichment are believed to be the most serious impact of SSOs discharging to small streams in Denmark. Such streams rapidly drain to either lakes or shallow coastal waters where eutrophication and long-term bioaccumulation are major problems.

## River Quality Classification for Transient Pollution

In order to produce a fully objective procedure for controlling pollution from SSOs, a classification framework is required within which the transient pollution resulting from SSO discharges may be related to the designated quality objectives of the receiving water. The present river quality classification system (NWC, 1978) is based around the compliance of routine monitoring samples with probabilistic river quality performance criteria. This approach does not effectively consider short duration pollution episodes, which may effectively limit the river biota, yet only occupy a very restricted time period.

The quality objectives for receiving waters are primarily related to the uses to which those waters are to be put. Uses may range from the supply of water for a variety of purposes, through fisheries and other biological aspects to amenity considerations. The significance of transient inputs of pollution will vary for each potential use and this must be reflected in the classification system that is developed.

A programme of research has been initiated at WRc Environment to investigate the relationship between acceptable concentrations of various pollutants for long- and short-term exposures and also the recurrence interval, or recovery period between exposure to various levels of pollutants. The end result is anticipated to be a classification matrix of criteria, related to the potential for use, pollutant concentration, duration and frequency for the major polluting determinands. This concept is presented in Figure 10.6 which is based

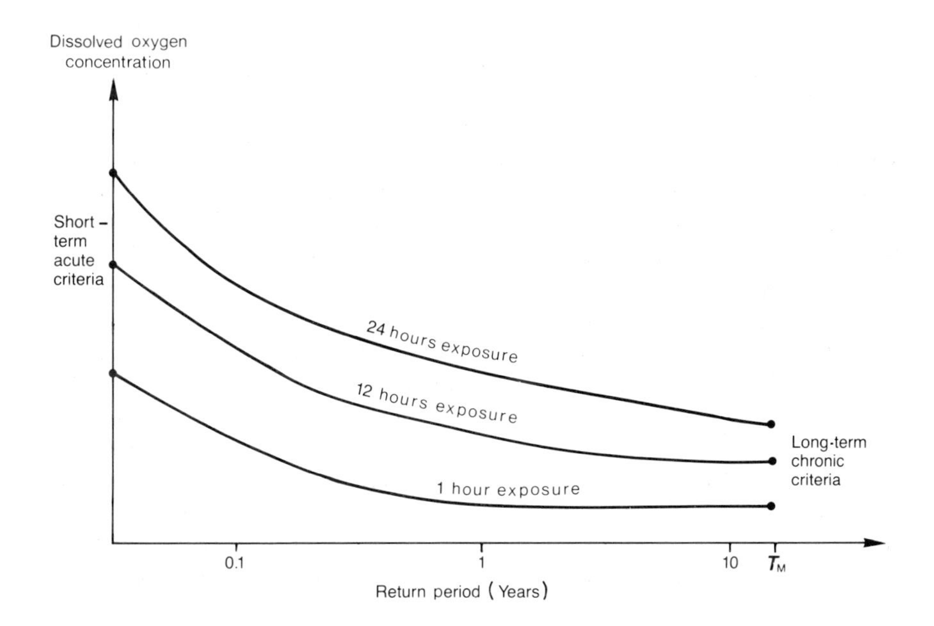


$T_M$ = minimum recurrence interval for an LC50 event to ensure survival of biological population

**Figure 10.6** Conceptual framework for an intermittent pollution river quality classification scheme. Reproduced by permission of the Water Research Centre

on the minimum dissolved oxygen levels permissible for a given exposure time and return period. The LC50, or lethal concentration to 50% of a biological population, is used as an acceptable performance indicator.

## INTERIM PROCEDURES FOR CONTROLLING SSO POLLUTION

### Implementation of SRM-II

The lack of a suitable sewer flow quality simulation model is a major problem in assessing SSO spill pollution load. The interim procedure proposes two simple methods to obtain an approximate load estimate. The first method is to take samples to characterize the base flow component and then use tabulated multiplying factors to incorporate the effects of storm runoff. This produces an average concentration which can be used to estimate the spill load associated with a volume of spill calculated using WASSP–SIM and TSR. The simpler second method is to use tabulated average storm sewage concentrations. As with the first method, these concentrations can be used to derive polluting loads from known volumes of spill. Both of these methods are extremely crude and offer a high degree of uncertainty.

### CARP — An Interim River Quality Planning Tool for SSO Spill Allocation

In 1983 North West Water published a public consultative paper (North West Water, 1983) setting out a strategy for dealing with water quality problems in its surface waters and estuaries over the next 25 years. While the proposed strategy has received general approval, the scale of the estimated costs is of concern to the Authority and to Government. A regional review of sewerage rehabilitation has been carried out to prepare more reliable cost estimates. The review involved the preparation of outline rehabilitation drainage area plans for seventeen sewer networks within the region. In review areas where SSOs are known to cause the receiving watercourse to fail its quality objective, the SRM-II interim procedure was used to calculate the spill load from each overflow. The lack of available river flow and time of travel data coupled with criteria to assess SSO pollution performance have restricted the development of a river quality planning procedure to cope with multiple SSO discharges to dendritic river systems. However, a procedure termed CARP (Comparative Acceptable River Pollution) was developed and was used, where appropriate, within the review.

The CARP procedure is used to calculate acceptable SSO spill loads to individual river reaches by comparing the pattern of load input rate per event with a reach which, whilst significantly affected, is not believed to be unacceptably polluted by SSO inputs, i.e. attains its designated quality objective despite SSO spills. The comparative statistic used is expressed in units of kg pollutant load spilled per event from all SSOs in a reach per km reach length per megalitre per day base flow. The use of a comparison of input rates is only valid if rivers with similar characteristics are compared and this assumes that acceptable spill loading rates implicitly include an assessment of the river's capacity for pollution assimilation and self-purification within each reach. CARP assumes that for each short time period transient event, the effects of the input to a reach are not passed down to the next reach. Hence the total river volume of the reach is available for dilution and mixing. This method produces sensible results but care should be taken when making interpretations of these results. Greater accuracy would be achieved if time of travel and pollutant decay rate data were available. The CARP approach avoids the need to specify recurrence frequencies for overflows and allows greater flexibility in design and planning by looking at individual reaches to identify where available capacity may exist in a river system or where the capacity is exceeded.

### Future Developments

The current WRc RBM programme, has been developed in response to an identified need of the UK water industry. The work in hand will, before about 1992, satisfy the immediate requirement for an objective methodology for the upgrading of sewerage systems in relation to river pollution from SSOs. In the longer term, this work can lead into the broader areas of river basin management and integrated river quality planning to include aspects of pollution control from other source types, such as agriculture and sewage treatment works.

### Acknowledgements

This chapter is published with the permission of the Director of the WRc. The views expressed are those of the author and not necessarily those of the WRc. The author would like to thank the other members of the RBM project team at WRc Engineering for their contributions to this work, in particular, Ian Clifforde, Rob Henderson and Paul Crockett.

## REFERENCES

Ackers, P. (1984). Sediment transport in sewers and the design implications, *Proceedings Conference Planning Construction, Maintenance and Operation of Sewer Systems*, BHRA, Cambridge.

Ackers, P., A. J. M. Harrison and A. J. Brewer (1968). The hydraulic design of storm sewage overflows incorporating storage, *J. Inst. Municipal Engrs*, **95**, 31–7.

Bowden, A. V. and Solbe, J. F. de L. G. (1986). Effects of intermittent pollution on rivers—derivation of water quality standards, *WRc ER 1227*, Medmenham.

Clifforde, I. T., A. J. Saul and J. M. Tyson (1986). Urban pollution of rivers—the UK water industry research programme. *Proceedings Conference Water Quality Modelling in the Inland Natural Environment*, BHRA, Bournemouth.

Crabtree, R. W. (1986). River water quality modelling in the UK. A review of past and present use with recommendations for future requirements, *WRc ER196E*, Swindon.

Crabtree, R. W. and Cluckie, I. D. (1987). Statistical methods for calculating effluent discharge consents, *Proc. Instn. Civ. Engrs.*, Part 2, Vol 81, 535–48.

Crabtree, R. W., I. D. Cluckie, C. F. Forster and C. P. Crockett (1987). Mathematical modelling requirements for river quality control, *Water Pollution Control*, **86**(1), 51–8.

Ellis, J. B. (1986). Hydrological controls of pollutant removal from highway surfaces, *Water Research*, **20**, 589–95.

Harremoes, P. L., (1982). Immediate and delayed oxygen depletion in rivers, *Water Research*, **16**, 1093–8.

Henderson, R. J. (1986). Rainfall time series for sewer system modelling, *WRc ER195E*, Swindon.

HMSO (1970). *Ministry of Housing and Local Government Technical Committee on Storm Overflows and the Disposal of Storm Sewage. Final Report*, HMSO, London.

HMSO (1977). Storm sewage separation and disposal. *Scottish Development Department*, HMSO, Edinburgh.

HMSO (1986). *River Quality in England and Wales 1985*, HMSO, London.

Hill, J. M. (1984). The acute toxicity of hydrogen sulphide—a literature review and application to storm sewage discharges, *WRc ER675M*, Medmenham.

Huber, W. C. (1984). *Storm Water Management Model, Users Manual Version III*. US EPA 600/2-84-109a.

Hvitved-Jacobsen, J. (1982). The impact of combined sewer overflows on the dissolved oxygen concentration of a river, *Water Research*, **16**, 1099–1105.

Johansen, N. B. (1983). Calculations of annual and extreme overflow events from combined sewers systems, *Int. Seminar on Rainfall as the Basis for Urban Runoff Design and Analysis*, Copenhagen.

Mance, G. (1981). The quality of urban storm discharges—a review, *WRc ER192-M*, Medmenham.

Matthews, P. J. (1987). COPA II: interpretation and application, *The Public Health Engineer*, **14**(5), 38–42.

National Water Council (1978). *River Water Quality—The Next Stage*. A Review of Discharge Consent Conditions, National Water Council, London.

National Water Council/Department of the Enviornment Standing Technical Committee (1981). Design and analysis of urban storm drainage—the Wallingford Procedure, *NWC/DoE Report No. 28*, National Water Council, London.

North West Water (1983). *Improving Rivers, Estuaries and Coastal Waters in the North West. A Consultation Paper*. North West Water, Warrington.

Pratt, C. J., G. E. P. Elliot and G. A. Fulcher (1986). Role of highway gullies in determining the water quality in separate storm sewers, *Urban Storm Water Quality and Effects upon Receiving Waters*, CHO-TNO, Wageningen, The Netherlands.

Price, R. K. and G. Mance (1978). A suspended solids model for stormwater runoff. In P. Helliwell (ed.) *Urban Storm Drainage*, Pentech Press, London.

Roswell, J. J. and R. Gaffoglio (1986). New York City—city wide combined sewer overflow study, *Proceedings 59th Annual Conference of the Water Pollution Control Federation*, Los Angeles.

Shuttleworth, A. J. (1986). State of rivers and sewers in Britain: is there a pollution problem, *The Public Health Engineer*, **14**(3), 49–51.

Stotz, G. and K. Krauth (1984). Factors affecting first flushes in combined sewers, *Proceedings 3rd International Conference on Urban Storm Drainage*, pp. 861–8. Gothenburg, Sweden.
Sonnen, M. B. (1977). *Abatement of Deposition and Scour in Sewers*, USEPA/600/2-77/121.
Thornton, R. C. and A. J. Saul (1986). Some quality characteristics of combined sewer flows, *The Public Health Engineer*, **14**(3), 35–9.
WRc/WAA (1983). *Sewerage Rehabilitation Manual*, WRc Engineering, Swindon.
WRc/WAA (1986). *Sewerage Rehabilitation Manual*, Second Edn, WRc Engineering, Swindon.

# Coastal Management

Geomorphology in Environmental Planning
Edited by J. M. Hooke



# 11 Geomorphology and Public Policy at the Coast

**ALAN CARR**
*Consultant, Wellington, Somerset*

## INTRODUCTION

It has been widely accepted that coastal management policies, including those with a geomorphological element, are disorganized and inconsistent (e.g. Gilg, 1978; Hansom, 1986). This stems partly from problems of definition; partly from the multiplicity of interests and organizations involved and their changing roles over time; and partly from the diverse, often short-term, pressures to which the coastline has been subjected during the post-war period. This chapter examines these various aspects mainly in the context of England and Wales. The first section looks at certain definitions, classifications and concepts. The next two sections concentrate largely on the kaleidoscopic effect of central government organizations and policies on coastal management since 1945. The penultimate part examines the fate of the geomorphological interest at a small number of specific sites, while the final section draws some overall conclusions from the experience of the last four decades. The thrust of this chapter differs from others in this book in that it is less concerned with actual and potential geomorphological input into government policy than in the way in which overall post-war governmental policy has affected an essentially passive coastal geomorphological milieu.

## THE WIDER CONTEXT

Great Britain is not unique (e.g. Hortig, 1972) in the varying ways that 'coast', 'coastal zone' or 'coastline' have been defined. For example, the Countryside Commission's Special Study Report *Nature Conservation at the Coast* (1969b) records that the National Parks Commission chose High Water Mark Mean Tide (HWMMT) to represent the 'coastline', with inlets generally being excluded; 'coastal belt' included all land one mile to landward of the arbitrary (HWMMT) datum. However, the Countryside Commission Report itself, while more or less keeping to the 'coastline' definition above, includes the inter-tidal zone and all estuaries and inlets. Furthermore, the one mile landward boundary has been replaced by a variable one which, in the case of 'soft' coasts, is often greater.

There are many ways in which the coastline can be classified: physically, strategically, or economically. It can be differentiated physically into such categories as cliffs or low coasts; rock or unconsolidated material. Such broad groupings may be subdivided, for example, into susceptibility to landslips or mudflows; or on the basis of geological composition. In the military context beaches may be assessed in difficulty of landing; how restricted access is away from the beach; and so on. Economically, criteria include scope for industrial development as exemplified by sheltered estuaries, deep water, and possible land reclamation. The beach and offshore zone may have potential as a source of aggregate for constructional purposes; such extraction can have repercussions in various ways. Indeed, one frequently recurring lesson is that it is not the initial development proposal, but rather implications deriving from it, that may be the greatest hazard. The coastline may also be classified as to the extent that it is built up, or remains in a natural or at least an agricultural state. Even within the field of recreation, land may be divided into active sports pursuits; scenic appreciation; or retirement bungalows or holiday chalets. There are car-parks, caravan sites and marinas which both provide, and of their own violation require, still further support facilities. Even in the narrow, scientific, sense, multi-disciplinary criteria have to be used in the assessment of values. This becomes still more true when other fields, such as socio-economics, are also taken into account.

Although this chapter will look at the coast and some of its problems primarily from a physical, essentially a geomorphological, point of view, there are some more fundamental questions that should be borne in mind. Is change necessarily harmful? What do we manage or conserve, assuming that conservation in one form or another is a desirable goal? Do we perpetuate a chance state in time? A dynamic artefact? Should we force our own interests and our own values on our descendants? In 1918 Carey and Oliver regarded the use of sand dunes for recreative 'joy lands' as being an unwarranted luxury. Even as recently as 1939 Dudley Stamp wrote of Dungeness: 'Unsightly as some of the wooden bungalows may be, this use of shingle areas may be commended. . . . Further, the preservation of more scenically attractive stretches of coastland becomes reasonable if such shingle areas are definitely made available as recreational tracts'. In the same paper he continued: 'The location of golf links on sand-dune areas is a good one; the links themselves are attractive . . .'. In conclusion Stamp wrote: 'If the needs of the country in the future are for land . . . then attention will of necessity be turned to such areas as Chichester Harbour, Langstone Harbour, as well as the great area of the Wash, which are available for reclamation'. In half a century ideas and values have fundamentally altered. Sand-dunes, and natural harbours such as Chichester, are now prized for recreation. Both dunes and shingle areas have conflicting roles such that scientific value and a desire for the status quo may vie with demands for development, building aggregate or public access. If such changes in value judgements and emphasis occur in the medium term it casts doubt over the degree of continuity that is likely in the future.

## PUBLIC POLICY: CHANGING AUTHORITIES AND DIVERSE ROLES

The words 'public', 'policy', 'coastal' and 'management' are subject to various interpretations and this is reflected, for example, in the different levels at which legislation and control are effected: parish, district, county, and national. In specific spatial or subject areas other criteria, such as those governing National Parks, or Water Authorities, may be relevant. However, even at central government level, the coast of England and Wales is susceptible to a multiplicity of different departments and interests. Furthermore, influences and control have changed with time. Thus the coast protection role of the Department of the Environment (DoE) was subsumed in 1985 with that of sea flooding under the Ministry of Agriculture, Fisheries and Food (MAFF). Similarly, in the field of nature conservation, the original 1947 proposals to set up a governmental body were devised under the aegis of the Ministry of Town and Country Planning Planning (later incorporated into DoE). They proposed that the conservation body should be linked with the Agricultural Research Council. This never happened. Instead the Nature Conservancy (NC) was established directly under the Privy Council. Later it became a component of the Natural Environment Research Council, and later still (as the Nature Conservancy Council (NCC)) found itself under DoE. While these arrangements are essentially administrative they also reflect changes in policy and in the ability to pursue an independent line.

The DoE continues to effect a direct interest in the coastal zone, not merely through the Department's planning responsibilities, and the NCC, but also in its role as a sponsor of certain aspects of research, and through the Property Services Agency (PSA), although it has recently been suggested that the latter might be privatized. Similarly, MAFF has additional roles other than in agriculture, food and sea defence. These include responsibility for monitoring radionuclides, such as those present in the effluent from the nuclear reprocessing plant at Sellafield (formerly Windscale), Cumbria. However, interests in the nuclear field—and all but one of the United Kingdom's commercial nuclear power stations are on the coast—is shared by the United Kingdom Atomic Energy Authority; the National Radiological Protection Laboratory; the Nuclear Installations Inspectorate; as well as the Atomic Weapons Research Establishment and British Nuclear Fuels. Again, all these bodies are government-sponsored or, in the last case, government-owned, while the Central Electricity Generating Board is, itself, a statutory body.

Finally, still at central government level, there are the research councils. Three of these, the Economic and Social Research Council, the Natural Environment Research Council and the Science and Engineering Research Council, all have a greater or lesser interest in the coastal scene.

While the title of this chapter refers to 'public policies' it is pertinent to remember that there are a whole range of semi-public bodies and organizations with an interest in the coast. Paramount amongst these is the National Trust (NT) which controls 760 km of the England and Wales coastline. This represents slightly over half the coast which the Trust deems in need of protection and worthy of such status. (The National Trust for Scotland only controls 2.5 per cent of the coastline there, reflecting the very long shoreline and the smaller perceived threat since some three-quarters of the Scottish Coast is classified as 'Preferred Conservation Zones' and much is in the hands of large landowners.) Because the National Trust for England and Wales' management policy has to pay at least some regard to the economic consequences, solutions to problems may be those of the land agent rather than the environmental scientist. Nevertheless the NT is unique in having inalienable rights to the land that it has acquired although how much protection such status would provide in a national state of emergency is open to doubt. Other bodies with a stake in the coast include the Council for the Preservation of Rural England; the Royal Society for the Protection of Birds (RSPB): and various Naturalists' Trusts. These organizations tend to concentrate on the aesthetic or the biological, rather than the geomorphological. Nevertheless they represent important elements in any equation which attempts to establish a form of balance as to the uses and abuses which may affect any length of coastline. As in the case of government organizations, this list is not intended to be exhaustive, but rather to demonstrate the fragmentation that exists and the diversity of points of view.

## THE CONCEPTS OF COASTAL CONSERVATION AND PLANNING

Virtually all laws relating to the coast fall under one of three broad categories; environmental conservation; sea defence; and planning. This section of the chapter successively reviews, mainly post-war, legislation in each area. However, it is impossible to separate entirely the laws concerning one subject area from their effects in another largely allied field. Additionally, other interests and aspects, such as transport and military defence, have major implications in the final outcome.

The government report *Conservation of Nature in England and Wales*, 'Cmmd 7122', was published in 1947. It outlined the state of conservation at the time and produced proposals for a, primarily biological, government organization to cover the subject. After four decades it is timely to re-read some of the committee's views.

> Most changes in nature are slow, insidious and not readily detectable; and they are often irreversible. An action which in itself appears sensible and desirable may have far-reaching and most unpleasant consequences, not forseen and possibly not appreciated for fifty years (p. 2).

> In spite of its dense population, this country still contains many places of the greatest scientific interest and not a few sites of classical importance in the history of knowledge. But these places are disappearing and deteriorating at an alarming rate. It is a depressing exercise to examine the Rothschild list of 1915 in the light of the condition of those same sites only thirty years later. Some have been irreparably destroyed, others are well on the way to destruction, and more have so far declined that they can no longer be rated as of outstanding national importance. This process is still continuing. Many could certainly have been saved had there been an active, informed, and centrally directed conservation policy; others are the inevitable consequences of two wars (p. 12).

The Committee proposed to include the conservation of wildlife and natural features as a 'planning purpose'.

> The reasons for safeguarding geological and physiographical features are not widely appreciated. In the conservation of these special areas for scientific purposes it is not expected that for most of them particular measures will be necessary (p. 24).

The Committee recognized that the coast presented special problems on account of its popularity and its narrowness as a zone. They believed that:

> A comprehensive national policy that will deal with . . . physical and engineering problems is clearly necessary; but it must be part of a wider policy also dealing with scientific and amenity factors (p. 27).

Returning to more general matters the Committee wrote:

> With regard to Governmental Departments no further or special action seems called for other than . . . liaison (p. 33).

While, specifically in the field of geomorphology (physiography) they observed:

> Direct Government interest in physiographical research is practically non-existent, yet the knowledge derived from such studies is of the highest importance in the practical solution of the many problems presented by coastal erosion and land-building, by the silting up of maritime and inland waterways (p. 51).

And on page 76 physiography was described as 'this most neglected branch of science' where 'systematic development of research . . . would save the wasteful expenditure of large sums of money'.

Many of the comments above are equally applicable today, although with hindsight one would have been less sanguine in the ability to devise a national policy; or for the safety of sites under governmental control; or, yet again, the ability for inter-departmental consultation to reach a satisfactory long-term result in the preservation of the natural environment against short-term pressures. Although we may be 'prone to extrapolate further unhindered activities to their surmised and fearful consequences' (Ippen, 1972) it is not necessarily true, as Ippen alleges, that 'the basis of such predictions is often more emotionally founded than established by adequate knowledge'.

The year 1949 saw two major pieces of legislation relevant to the coast. These were the National Parks and Access to the Countryside Act, and the Coast Protection Act. Part I of the former outlined its purpose:

> for preservation and enhancement of natural beauty in England and Wales, and particularly in the areas designated under this Act as National Parks or areas of outstanding beauty

It pointed out that the National Parks Commission (later to become the Countryside Commission) was not to be a planning authority. An Area of Outstanding Natural Beauty was defined as any area in England and Wales not in a National Park, which appears to the National Park Commission 'to be of such outstanding natural beauty . . . that the provisions of the Act . . . should apply thereto . . .'.

In 1963 the then Ministry of Housing and Local Government sent out a circular to all local planning authorities with coastal boundaries for 'action'. Under the heading 'Coastal Preservation and Development' the document stated:

> There is a need to give increasing attention to the problem of coastal development. The coast attracts increasing numbers of people for holidays and week-end recreation. The same is true of retirement. It is most important that in making provision for these needs that the development permitted should not spoil the very things which give the coastline its charm and attraction. In addition to its high amenity value the coastline contains many areas of great scientific interest by virtue not only of their natural flora and fauna but also of the physiographic features. Many of these are unique and irreplaceable so that if they were lost the richness of the scientific interest of Britain's coastline would be permanently impoverished.
>
> Circular No. 56/63 dated 2 September 1963.

Both this circular and Cmmd 7122 assume an integrated policy in coastal planning and, furthermore, both give substantial recognition to 'physiography', i.e. geomorphology, as a relevant science and a type of feature worthy of preservation. It seems that, at least in the coastal sphere, this mention of geomorphology was never fully acted upon either in the formulation of overall policy nor in its detailed implementation. To what extent this was the fault of the planners (whose geographical background and hence awareness of geomorphology has diminished over time), and to what extent a lack of relevant skills and proselytizing zeal by coastal geomorphologists (Carr, 1987) may be a matter of discussion.

The period around 1970 was marked by conspicuous activity by the Countryside Commission in the field of coastal preservation and development (Countryside Commission, 1968; 1969a,b; 1970a,b). Probably the most significant of their reports was *The Planning of the Coastline* produced on behalf of the (then) Ministry of Housing and Local Government and the Welsh Office. In its introduction it states:

> For the most part, these pressures for development (in leisure, dock and port facilities, steel mills, oil refineries, nuclear and oil fired power stations and other demands 'as yet unknown') are in conflict with the interest in natural history and scientific research.

The report is mainly concerned with planning and administrative organization. Nevertheless there are a number of important general and specific recommendations. These include: the stopping of the arrangement whereby land acquired for defence purposes in wartime was transferred to another government department for entirely different purposes thereafter; and the designation of 34 areas, totalling approximately 1200 km in length, as 'Heritage Coast' where development should be severely restricted. The report observes that:

> Over much of the coast the protective policies of the local planning authorities are so vague that it is impossible to be sure what they mean in practice.

Roughly contemporaneous with this period of activity was the United States' Coastal Zone Management Act of 1972. Although inadequately funded and described as a 'timid, reluctant and . . . contradictory first step forward' (Stahr, 1974), this federal legislation in part reflected existing environmental concerns, and in part generated further interest in the coastal zone. While the response of individual maritime states varied, the legislation had the overall effect of increasing environmental awareness. This is reflected, for example, in the popular scientific writing related to the coast. The latter differs from the UK experience, in that it tends to be by well-known and well-qualified specialists (e.g. Pilkey *et al.*, 1978).

One further item of legislation in the field of conservation needs to be mentioned. This is the Wildlife and Countryside Act of 1981, which has had an equivocal response. It modified the National Parks and Access to the Countryside Act, 1949, which provided most of the Nature Conservancy's original powers. The principal merit of the Act is that it provides some security for geomorphological and other sites which would be endangered by activities such as agricultural improvement or forestry. Providing the NCC has notified an interest in a site then they must be consulted even if proposals do not constitute development and, hence, do not require planning permission. However, the Act required the NCC to re-schedule all Sites of Special Scientific Importance (SSSIs) and also imposed an obligation to provide financial compensation.

The second strand in coastal legislation is that of sea defence, where until quite recently there were two separate lines of activity. These related to the destruction of property and erosion (coast protection); and sea flooding. The 1949 Coast Protection Act was defined as:

> An Act to amend the law relating to the protection of the coast of Great Britain against erosion and encroachment by the sea.

The Act proposed that:

> The council of each maritime county borough or county district shall . . . be the coast protection authority.

The Minister was empowered to create a Board comprising one or more councils, river boards, drainage or highway authorities, and the British Transport Commission, as considered appropriate. These arrangements were to be co-ordinated by the Ministry of Housing and Local Government, later to become part of DoE. Analogous provisions applied to Scotland. The Act allowed any coast protection authority to carry out 'such protection work, whether within or outside their area, as may appear to them to be necessary or expedient for the protection of any land within their area'.

Section 18 of the Act provided that:

> it shall be unlawful to excavate or remove any materials . . . on, under, or forming any portion of the seashore.

unless the Coast Protection Authority had given a licence. However, different provisions applied to firms already extracting prior to the Act. In those instances it had to be demonstrated that the effect was harmful, an almost impossible task. The Act gave some rights to the aggrieved party to obtain compensation. Section 18 did not apply to Scotland.

The 1954 Report of the Departmental Committee on Coastal Flooding (the 'Waverley Committee' set up in response to the East Coast floods of the previous year) expressed the views, firstly that the responsibility for coast defence should not be placed on central government and, secondly, that the existing division between the river boards with their concern for drainage and flooding, and local government coast protection authorities responsible to the MHLG, was both clear and logical.

The Land Drainage Act dates from 1976 and has been considered as complementary to the 1949 Coast Protection Act. Whereas the 1949 Act was concerned with powers to construct works to prevent erosion and encroachment the 1976 Act involved legislation to protect land against sea flooding. Unlike the Coast Protection Act the Land Drainage Act does not refer to Scotland where legislation dating from 1958 and 1961 is operative. In 1985, as noted earlier, the situation was rationalized with MAFF taking sole responsibility for both coast protection and sea flooding. However, anachronisms remain with local authorities being sub-contracted to carry out supervision of sea defence projects as their grant rate may be higher than that of water authorities. This situation has applied at Chesil Beach, Dorset (see below).

In Japan, where national disasters were far more serious, legislation passed in 1956 empowered the state to adopt uniform policies along the whole coast (Moyazaki, 1969).

Some 565 km (20 per cent) of the English coastline is protected against erosion; an almost identical amount (560 km) is protected against sea flooding (Herlihy, 1982; Trafford and Braybrooks, 1983). Cost of capital works is similarly divided. Thus in 1980/81 coast defence totalled some £16 m plus £2 m for maintenance; capital works for sea flooding measures came to £17 m. One of the features of importance is the way in which sea defence schemes are very unevently distributed around the English and Welsh coasts with nearly half the works related to sea flooding falling within the area of the Anglian Water Authority (Carr, 1984). This is the effect of the large proportion of low-lying land; relatively rising sea level in the area; and susceptibility to storm surges. Concomitant with this is the higher level of grants from central government (up to 85 per cent) relative to those available to other authorities.

Cost benefit analysis was first considered by DoE in 1978. This concept is discussed further in a paper by Parker and Penning-Rowsell (1983), and also later in this chapter in the specific context of Chesil Beach. It is sufficient here to say that even if a particular sea defence scheme may not be warranted on economic grounds it may be justified in an attempt to treat an area as an entity rather than as isolated components.

The legislation discussed in this section of the chapter so far has been almost entirely concerned with 'conservation' in its broadest sense. There is one major area to which reference has been made only in incidentally; the law related to town and country planning. In the period between 1943 and 1981 there were no less than fourteen Acts with the words Town and Country in their titles (Williams, 1981). There were a further seven Acts in the immediate planning field quite apart from the various pieces of environmental legislation referred to earlier. While many of the planning Acts do not involve the coast, or do so only indirectly, they demonstrate the instability of planning as a whole. Thus each of the Acts of 1947, 1962, and 1971 repealed all previous planning law, while (for example) legislation concerning compensation for loss of development rights enacted by Labour governments in 1947 and 1975 was repealed by Conservative governments in 1953 and 1980, respectively. Many aspects of planning which were undertaken at county level until the Local Government Act of 1972 were devolved to district level thereafter; these responsibilities were redefined in further legislation in 1980.

## EXPERIENCE AT SPECIFIC SITES

Partly because of the writer's interest in, and knowledge of, shingle beaches, and partly because England was uniquely endowed with several examples of international importance, this section concentrates mainly upon such features: Orfordness, Suffolk; Chesil Beach, Dorset; and Dungeness, Kent (Figure 11.1). The geomorphological interest of the first two sites is the paramount reason for their being scheduled for protection; in the case of Dungeness it is a very significant factor. Other locations, of different composition and situation, are discussed mainly to see to what extent it is possible to generalize. We are less concerned here with the places as such; rather more with the extent to which they demonstrate the influence of different policies and pressures.

Dungeness is the most prominent feature on the 40 km long shingle coastline of Sussex and Kent. The 2200 ha cuspate foreland consists of some 500 ridges in apposition one to another, the oldest of which are thought to date from as early as 3400 BP (Eddison, 1983; Henderson, 1985). Orfordness (Figure 11.2) extends in a south and south-westerly direction for some 15 km from Aldeburgh. It may be divided into three components: the coastline north of the Ness proper which has been steadily eroding for many centuries; the cuspate foreland of the Ness; and the shingle spit which terminates in the estuary of the Ore–Alde river system. As in the case of Dungeness, the structure of the whole shows (or in this instances more correctly showed) groups of ridges with different crest levels. It is thought that these systematic variations may represent changes in relative sea level, or wave energy, over time.

The third shingle feature, Chesil Beach (Figure 11.3), is of a fundamentally different, linear form (Carr and Blackley, 1974). It extends from the so-called Isle of Portland in the east some 18 km or more to an essentially arbitrary limit in the west. For the present chapter this will be taken as West Bay, Bridport, in accordance with local usage. For much of its length the beach is backed by the shallow, tidal, Fleet lagoon. The international importance of Chesil is due to its sheer size, reaching nearly 14 m above mean sea level in the east; its geological composition; and the systematic longshore size-grading of the pebbles and cobbles.

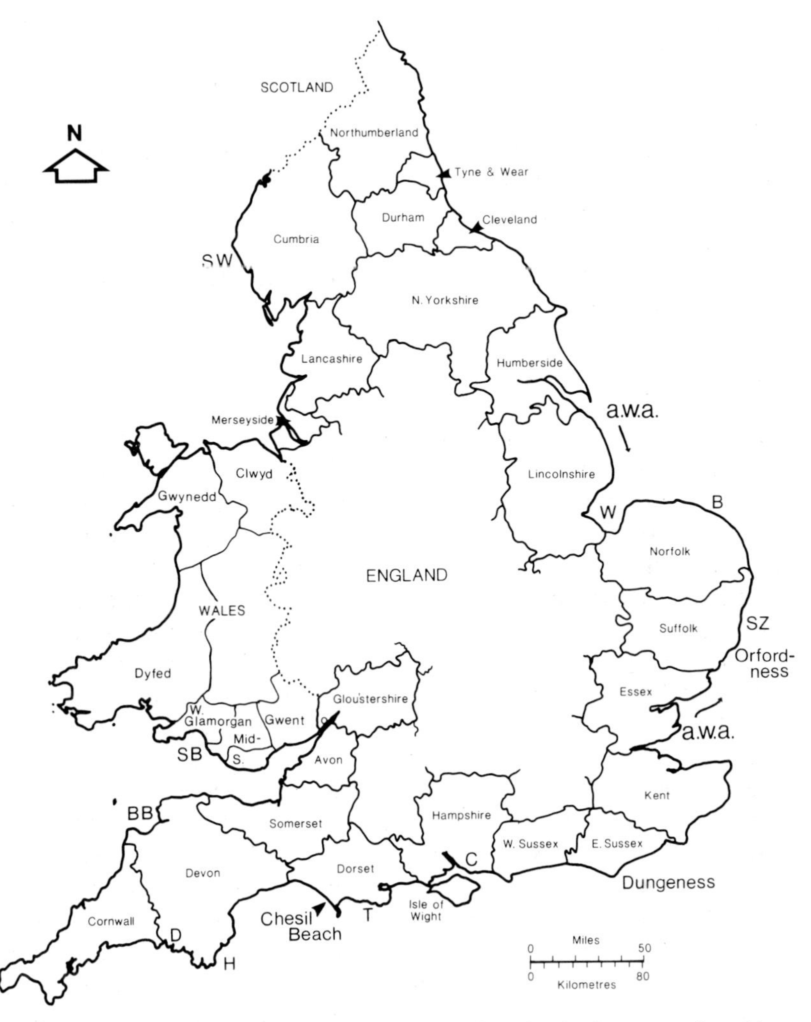


**Figure 11.1** England and Wales, showing main places referred to in the text and maritime county boundaries effective from April 1974. The N and S limits of the Anglian Water Authority (a.w.a.) are shown. B = Bacton; BB = Braunton Burrows/Taw–Torridge estuary; C = Chichester/Langstone harbours; D = Devonport docks, Plymouth; H = Hallsands, Start Bay; SB = Swansea Bay; SW = Sellafield/Windscale, Cumbria; SZ = Sizewell; T = Tyneham/Worbarrow; W = Wash

## Orfordness

Carr (1986) has listed the various papers which have been written concerning aspects of the area's geomorphology. Kinsey (1981) has described the military use of Orfordness from the beginning of the First World War until 1981. Although there was considerable activity at various times between 1915 and the early 1950s such activity was concentrated either on the King's Marshes

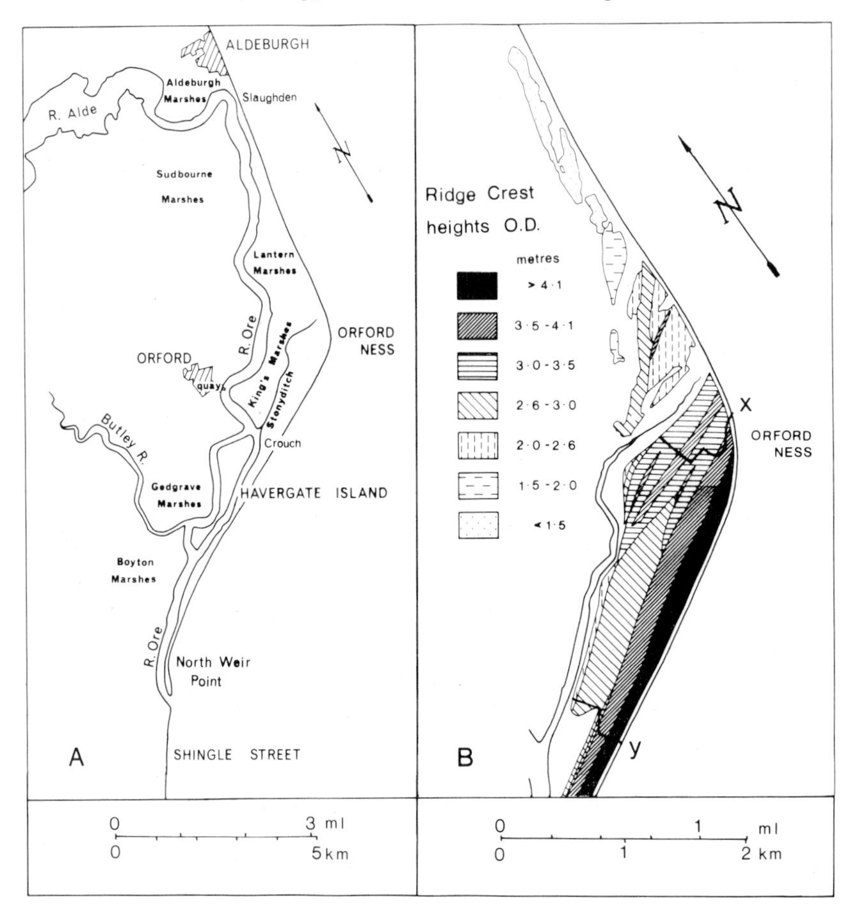


**Figure 11.2** Orfordness: (a) Site map. (b) Ridge height diagram; area N of 'x' totally destroyed 1965 onwards; area N of 'y' partly destroyed 1953 onwards

(i.e. the southern reclaimed marshes) or on the target area of shingle immediately north of Stonyditch creek. While this spread was extensively pock-marked it was still possible to recognize the ridge structure and to obtain height measurements from undisturbed parts in the early 1960s. Air photography in early 1953 shows the whole suite of ridges from immediately south of Slaughden, Aldeburgh, to the distal end of the spit as largely intact. Apart from access tracks to the lighthouse and small isolated buildings (such as that used by Watson Watt for his initial experiments in radio direction finding in 1935) little disturbance of the ridge structure has occurred. Aerial photographs taken in 1959 show three areas of new buildings and a quite substantial road network. All this new development, largely related to the atomic weapons development programme, was situated on the shingle structures of the Ness and was not subject to planning controls. Although the buildings were themselves substantial the adjacent areas of disturbance and ridge destruction cover more than ten times as large an area. Aerial photographs taken in 1964 record two new pagoda-like laboratories constructed in 1960–61. In this case, partly because shingle has been banked up against the buildings, the ridge structure has been eliminated from an area about twenty times that of the buildings themselves. (Subsequent 'restoration' of the derelict laboratories on aesthetic grounds would be likely to make the situation still worse.) The Atomic Weapons Research Station closed in 1971.

Further damage to the shingle structures, in a hitherto unaffected area adjacent to Stonyditch, took place in 1965 with the removal by light railway of 205 000 $m^3$ of material to feed the eroding beach at Slaughden. Construction of the abortive 'Cobra Mist' advanced warning radar system at Lantern Marshes began in late 1967. The project was abandoned in 1973 after just two years in actual operation. It resulted in the elimination of all the shingle structures north of Stonyditch. Some further disturbance of a relatively minor nature occurred during the demolition of unexploded weapons in 1967 and again between 1973 and 1986. Routine maintenance of sea defences has had similar-scale effects.

The record of the last three decades is a sorry one. Nor does it look as if the future, with continuing expressions of interest from, for example, the CEGB is likely to be substantially better than in the recent past. Orfordness has been tentatively suggested as the site for Britain's first fusion reactor power station on grounds of its remoteness. The *East Anglian Daily Times* of 16 September 1982 described a 1979 document which suggested the construction of up to six nuclear power stations on Orford Ness, out of a total of nine in Suffolk. The county council's chief planning officer stated that any proposal to build a power station at Orfordness would be 'unacceptable use of part of the Heritage Coast'. However, such designation is unlikely to be proof against the perceived 'national interest'.

While over the post-1945 period attempts to preserve the biological and geomorphological significance of parts of the Orfordness–Havergate NNR and SSSI have proved successful, major episodes of the geomorphological record have been destroyed for very transitory benefits in the field of defence and short-term local employment.

### Chesil Beach

There is a marked contrast between the ends of Chesil Beach, with their greater economic importance and their relative dearth of beach material, and the less problematic central area. This is reflected in aspects of management in response to the various pressures. The central portion, with the associated Fleet lagoon, belongs in large measure to a single estate. To the extent that damage to the environment has occurred there it has been due almost entirely to peripheral pressures, such as nearby caravan sites and resulting problems of sewage

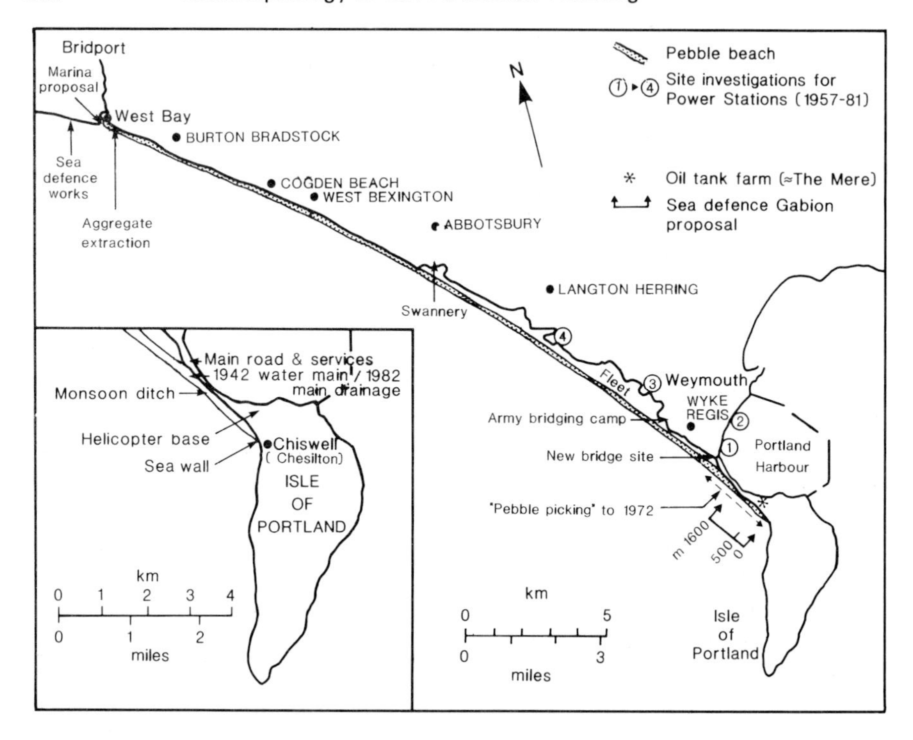


**Figure 11.3** Chesil Beach: site map. The former bulk aggregate and 1906 extraction sites; existing caravan sites and car parks are omitted but most other significant developments are indicated. The 1980 'Monsoon ditch' was replaced in 1986 by a larger interceptor channel also draining into Portland Harbour

disposal. Figure 11.3 shows sites investigated for power stations, the most recent two of which would have been nuclear and would have fallen into this central area. The East Fleet site [3] would have resulted in a fundamental change in the hydraulic regime of the lagoon.

Bulk extraction of aggregate has taken place at a number of locations along the foreshore of the western portion of Chesil Beach for centuries and only finally ceased in 1987. Initially the rate is likely to have been low but Carr (1983) suggests that between the mid-1930s and 1977 about one million tonnes (i.e. some 2 per cent of the whole beach) had been removed. Of the sites still in operation after 1945 probably the most interesting is that at West Bay, Bridport, where longshore transport of pebbles and sand is impeded by the harbour jetties. This situation began in the mid-nineteenth century when the piers were filled in to reduce sedimentation at the harbour mouth. The net result was that the exposed western side became starved of beach material while the relatively sheltered eastern side suffered from an excess. The corollary was that, until 1984, sea defence works were undertaken to the west by the Wessex Water Authority with aid from central government funds while extraction took place immediately to the east with the local authority obtaining a royalty from the contractors. This anomaly ceased when further licences were refused, but it is a strong argument for both sea defence works all falling under one ministry as is now the case, and also for the local authority having some financial stake in the work.

The small settlement of Chiswell, nestling under the cliffs of the Isle of Portland, has probably always been under some threat of flooding by waves through seepage and over-topping of the beach crest, thus buildings were constructed to minimize the consequences. There has been argument as to whether the incidence of flooding and its severity has varied over time. One factor that may have a bearing in this is the removal of substantial quantities of pebbles from the backslope of Chesil Beach just north of Chiswell in 1906 in order to provide the foundations for the nearby oil tank farm. Such removal is a demonstration of expediency and it closely followed the extraction of some 395 000 $m^3$ at Hallsands, Start Bay, to extend Devonport docks (Hails, 1975). In the case of Hallsands the consequences were almost immediate with the initial impact just a year following completion of the project and the final demise of the village only sixteen years later, in 1917. Cause and effect is less immediate and less obvious at Chesil but is probably just as real. Another factor in the increased tendency to flooding is the infilling of the sandy area of the Mere which faced on to Portland Harbour. This was undertaken in 1962–3 to provide a helicopter base. While some provision for drainage was made, the net result was that water reaching Chiswell under storm conditions was no longer able to drain freely into Portland Harbour.

To reduce the incidence of flooding in Chiswell a sea wall was constructed in 1962 and extended in 1965. Quite apart from the question of the appropriateness of solid defences at the particular location, the wall did not extend sufficiently far to protect the settlement as a whole, while there were problems at the junction of the sea wall and the natural beach. Furthermore, perhaps because of the weak underlying geological strata, the sea wall was under-designed and liable to be over-topped under precisely those conditions it was intended to guard against.

Lewis (1979) gave the population of Chiswell in 1942 as 134. This low figure partly reflects the effect of flooding and bombing during the 1939–45 war but the immediate post-war policy of the local authority to demolish housing which it regarded as sub-standard meant that the figure remained low. It was also influenced by the planning blight caused by the proposal, later abandoned, to build a new road through the area. Lewis believed that it would have been possible to raze the village at that time and move the relatively few inhabitants into new dwellings elsewhere. Instead, new and inappropriate construction was carried out to provide a 'seaside character' and, with the coming of highly articulate people from outside the immediate area, pressure was put upon the local authority to upgrade the infrastructure. Chiswell became a General

Improvement Area (GIA) in 1982, having been designated a 'Conservation Area' in 1975. GIA status entitled it to a range of grants to enhance civic amenities. While the incidence of flooding may have been little different from that previously, the material consequences were more severe. However, the justification for the programme of sea defences, costed at £4.5 m plus maintenance in 1980, was based largely on the economic and strategic consequences to the Isle of Portland rather than those to Chiswell alone. Given the range of assumptions, Penning-Rowsell and Parker (1980) suggested that the proposed scheme would bring assessed benefits of £12.1 m. There would have been other options, however, and these might have been more scientifically and environmentally acceptable. For example, the settlement at Chiswell might have been protected by raising the height of the sea wall as has now been done, but in conjunction with a return embankment immediately north of the village. In complementary fashion, the main access route to the Isle of Portland could then simply have been raised and its seaward face armoured. This combined option would have meant that only a relatively short stretch of Chesil Beach would have been damaged by an extension to the existing sea wall or through gabions and mattresses. It is relevant to add in this context that the wire mesh gabions and mattresses while probably more acceptable aesthetically are not necessarily scientifically so. Apart from greater maintenance requirements this is because the cages in the 500 m trial section were either filled with limestone fragments of local origin, or graded beach material. Thus the unique size-grading of the beach, which until the 1973 Public Inquiry had been threatened by selective pebble-picking from opposite Portland Harbour to Chiswell was now modified instead by the requirement to fill up the cages with specific size fractions. In addition, the plan to extend the trial area a further 1100 m towards the northwest would entail the import of foreign geological constituents since paradoxically there is inadequate suitable local limestone available.

In summary, Chesil Beach, especially the eastern end, is a good example of changing policy and pressures. Over the post-war period a range of alternatives has been available from time to time. The decisions ultimately taken reflect political expediency. There were other simpler, cheaper solutions had the problems been taken in time and various points where the scientific interest could have been better served. Decisions have been taken in isolation. One thing that Chesil, in common with Orfordness and other sites lacked, and continues to lack, is a consistent long-term policy. Not only is planning unfortunately 'an evolving process' (Williams, 1981) but it is also true that '. . . public opinion is no less changeable than technology' (Eddington and Eddington, 1977).

### Dungeness

Dungeness, like Chesil Beach, has been regarded as of international importance, and it, too, has been described as 'unique'. The cuspate foreland has already been mentioned in the context of Stamp's (1939) low opinion of its scientific value. It may well be partly this evaluation that made it so difficult for the Nature Conservancy to defend the site during the 1958 Public Inquiry prior to the construction of the first nuclear power station there. Other factors were the dismissive attitude of the media at that time to environmental causes, and that the Inspector appointed by the Ministry of Fuel and Power was closely allied to the electrical generating industry. (The particular expertise of inspectors chosen for Public Inquiries has frequently presented problems for environmentalists. Thus sea defence schemes have often been evaluated solely by a civil engineer; the impact of sand and gravel winning by someone well-versed in town planning law but perhaps unable to fully appreciate the scientific implications.) It was also hypocritical to suggest in the case of Dungeness that the cause of conservation would be able to be reconsidered when planning permission was sought for the power lines from the station. Once the generating station was approved it was inconceivable that the transmission lines would not be.

It was pointed out at the Public Inquiry that the proposed site was on an eroding shoreline. Nevertheless this evidence was ignored with the result that almost constant beach nourishment has taken place since the construction of the initial, 'A', station. This feeding consists of about 30 000 m$^3$ per year, with a similar amount nourishing the local sea defences. According to Eddison (1983) the total exceeded 110 000 m$^3$ in 1979. Initially the shingle used for beach replenishment was extracted from an area of biological importance.

Findon (1985) and Fuller (1985) have reviewed the nature of the pressures, and the resulting effect on the shingle structure at Dungeness, respectively. Findon pointed out that most of the pressures date from after the mid-nineteenth century, although rather earlier than those on Orfordness and Chesil Beach. The western part of Dungeness, Lydd Ranges, was first occupied by the army about 1880, and was used for heavy artillery; then tanks; and finally for small arms practice. There are at present some 0.5 million man training days per year on the ranges, virtually all of which fall within the scheduled SSSI. Most of the environmental damage on the ranges occurred by 1945; Fuller suggests that only 4 per cent of the ridges are now essentially unharmed.

The first aggregate extraction, and the arrival of the railway, both date from 1883. According to Fuller (1985) 420 ha (19 per cent) of Dungeness has been affected by gravel extraction, with 170 ha having been removed to below the water table. Findon (1985) notes that up until 1976 planning applications covered 726 ha while peak extraction was reached in 1973 with 1.5 m tonnes of material being removed. Since 1974 production has averaged about 1 m tonnes, which represents rather more than 10 ha each year. The RSPB is one of the principal landowners at Dungeness. According to Scott (1985) extraction from the RSPB reserve began in 1970 and ended in 1977; at its peak it was at the rate of 535 000 m$^3$ of pebbles and 99 000 m$^3$ of sand a year. While Scott refers to the 'lengthy discussions' carried out with other conservation interests, it is clear that ornithology and commercial considerations totally dominated the final outcome.

Other pressures include water extraction and construction work. While a scheme for building storage reservoirs lined with butyl rubber was abandoned, vehicle access to wells located within the shingle foreland is a major problem. Tracks are not subject to planning constraints; they may also be caused by such factors as line maintenance from the power stations, and leisure activities. Fuller states that housing on the eastern shore doubled between 1946 and 1984. Of the 5 per cent of Dungeness built up only 2 per cent (i.e. two-fifths) represents the CEGB site, but the land acquired by the board comes to 6 per cent in total. Construction of Dungeness 'C', proposed in 1974, would entail a further 30 ha most of which would be in one of the best remaining ridge areas.

As in the case of Orfordness it is not solely the actual building works that are of significance but also the ancillary works and their implications. At the 1958 Public Inquiry the Nature Conservancy stated that if the power station went ahead the entire site would be written off in conservation terms. Even geomorphologically this was not entirely true, and was a view that was to be regretted subsequently. According to Fuller (1985) the period between 1946 and 1958 saw virtually no change at Dungeness. The NC decision to abandon the National Nature Reserve proposals gave the 'green light' to further development. In 1958 41 per cent of the shingle ridges were intact; this figure had dropped to 28 per cent by 1984. The Nature Conservancy reconsidered their policy in 1968 and part of Dungeness was re-designated as a Proposed National Nature Reserve in 1970.

Dungeness, like Orfordness, brings one back to the question of residual value; the premium placed on remoteness for certain types of engineering activity; and the progressively increasing ability to construct major works inspite of adverse ground conditions. While this may incur a financial penalty this is only rarely paid directly by the developers. Two further points should also be mentioned in the context of Dungeness. Firstly, there is the total elimination of all features, geomorphological and otherwise, that takes place in the large part of the foreland subject to mineral extraction to below the water table. While this may be understandable in a *commercial* context, it is less easy to accept the attitude of the RSPB, which in the furtherance of its ornithological interests and the acquisition of revenue, has actively participated in such severe environmental damage to the area which they owned. It would seem that even in the limited realm of conservation it is impossible to achieve any coherent policy. Secondly, as Findon (1985) noted; 'The vast majority of the pressures on Dungeness lie outside the scope of the Wildlife and Countryside Act, 1981, because they are planning issues where the District and the County Council or indeed Central Government will be the final arbiters'. '. . . planning decisions are made by elected representatives who have a less than perfect understanding of the issues and the international importance of Dungeness. . . .'

Not all development schemes come to fruition, whether because of physical and technological factors; economic constraints; or vagaries in the political climate. Examples include, in the case of Dungeness, the proposal submitted to the National Ports Council, for the development of a major deep water port in the lee of the foreland (Anon, 1971). Port Dungeness would have involved 2800 ha of reclamation and 1330 ha of development on the shingle area of the foreland. Better known examples are the initial investigations as to the feasibility of a third London airport at Maplin, Essex—cancelled by a change in central government—and various barrage schemes, such as that for the Wash. Other proposals that did materialize include the Bacton gas terminal on the north Norfolk coast, and numerous oil-related projects such as Sullom Voe, Shetlands, and Strathbeg and Nigg Bay on the Scottish mainland. Both oil and gas-related works would not have been anticipated until about the 1960s and the scale was not recognized until a decade or so later. While, as Davis (1957) said in another context: 'Management . . . amounts to foresight . . .' it is often impossible to anticipate the direction from which new threats and opportunities may come.

While geomorphological interest rarely, if ever, benefits from the concept of residual value, there are a few occasions where pressures on the natural environment may not be irrevocable. Such was the case at Braunton Burrows, north Devon. This dune system, widely regarded as the best in the Southwest, was subject to intense military use which stripped the vegetation, especially from the central third, both during the 1939–45 war and immediately thereafter. Some, relatively limited, training still continues. However, the wartime use, and post-war attempts to remove mines by means of pressure hoses, meant that in the central zone the broad topographical pattern had become inverted with the pre-war ridges becoming the post-war slacks, and *vice versa* (Kidson and Carr, 1960). The area exceeding 100 feet (30 m) in height was multiplied thirty-six fold. These dramatic changes did not extend to the northern or southern thirds which retained most of their original form. In the Braunton case, the topographical changes which took place have not really done any significant harm to the geomorphological importance of the site, and might even have been capitalized upon for research purposes under different circumstances.

Both the Taw-Torridge estuary adjacent to Braunton Burrows, and Swansea Bay on the other side of the Bristol Channel, demonstrate the difficulty of proving that the erosion of the coastline was due to the nearby extraction of aggregate. In each instance the evidence provided by consultants and expert witnesses was contradictory, while in the case of Swansea Bay some £0.5 m of public money (at 1980 prices) was spent to prove that the beach was not fed from offshore. It is arguable that such expenditure should not come out of the public purse but should be provided by the direct beneficiaries. The eastern shore of Swansea Bay is also an example of conflicting policies of adjacent county councils. Spoil from the planned extension to the steel works at Margam, West Glamorgan, was tipped on the north bank of the River Kenfig, directly opposite the local nature reserve which falls in Mid-Glamorgan. The British Steel Corporation scheme was eventually abandoned with the contraction of the iron and steel industry in the late 1970s with the result that, in this instance, the local environment had been damaged without any economic benefit having been realized at all.

## CONCLUSIONS

Both the broader picture and the specific site examples point to a number of lessons and general precepts which affect the interrelation of geomorphology and public policy in the coastal environment. Many of the implications apply more widely. There are too many organizations at too many levels. Nor is there any coherent and consistent policy at any one time or over time. While some of these shortcomings are inevitable it is impossible to condone the majority.

In the US Stahr (1974) wrote: '. . . the national interest is not how much short-term gain we can extract from the coastal zone at the expense of long-term values'. The post-1945 record in Great Britain, especially in England and Wales, suggests that we have failed to recognize this fact. Furthermore, this failure is compounded in the sense that the worst examples are frequently those of government departments and statutory bodies supposedly acting in that same 'national interest'. Economic and political pressures, together with growing technological expertise, have resulted in increasing damage to the environment for only transitory benefits. This is particularly apparent in the case of Orfordness.

The last two decades have seen a decline in the role of geomorphology in the coastal field. This may be attributed partly to the changing background of personnel in the realm of planning. It may also be explained by the continuing biological dominance in conservation and of civil engineering in sea defence (and of both in pollution control), as well as the greater need for resources and relevant qualifications in research and site evaluation. While mathematical expertise and knowledge of physics is desirable, they frequently appear to diminish environmental awareness. Examples include the inappropriate parameters adopted by mathematical modellers to trace the fate of radioactive effluent from the reprocessing plant at Sellafield, Cumbria, and a comparable lack of understanding in proposals as to how to deal with the vessel *Elani V* after oil spills along the East Anglian coast in 1978.

Geomorphology has a role to play in the coastal field even if, currently, that view appears to be accepted more widely in other countries than Great Britain.

### Acknowledgements

Assistance in providing data and information for this chapter has come from many directions. In particular I would like to express my thanks to S. Morris;

E. Amos, R. Findon and C. J. D. Shackles (NCC); and G. M. West (WWA). The views and prejudices are entirely those of the author, however.

## REFERENCES

Anon. (1971). Dungeness has major port potential, *Dock Harb. Auth.*, **52**, 96–102.

Carey, A. E. and Oliver, F. W. (1918). *Tidal Lands*, Blackie and Sons, London.

Carr, A. P. (1983). Chesil Beach: environmental, economic and sociological pressures, *Geogr. J.*, **149**, 53–62.

Carr, A. P. (1984). England, pp. 181–8. In H. J. Walker (ed.) *Artificial Structures and Shorelines* (abbreviated version), Baton Rouge, USA.

Carr, A. P. (1986). The estuary of the River Ore, Suffolk; three decades of change in a longer-term context. *Field Studies*, **6**, 439–58.

Carr, A. P. (1987). The coastal papers in a broader context. (Introduction to section on coastal geomorphology.) In V. Gardiner (ed.) *International Geomorphology 1986*, Vol. 1, pp. 1115–17, Wiley, Chichester.

Carr, A. P. and M. W. L. Blackley (1974). Ideas on the origin and development of Chesil Beach, Dorset, *Proc. Dorset Nat. Hist. Archaeol. Soc.*, **95**, 9–17.

Countryside Commission (1968). *The Coasts of England and Wales: Measurement of Use, Protection and Development*, HMSO, London.

Countryside Commission (1969a). *Coastal Recreation and Holidays*. Special Study Report Vol. 1, HMSO, London.

Countryside Commission (1969b). *Nature Conservation at the Coast*. Special Study Report Vol. 2, HMSO, London.

Countryside Commission (1970a). *The Planning of the Coastline*, HMSO, London.

Countryside Commission (1970b). *The Coastal Heritage—a Conservation Policy for Coasts of High Quality Scenery*. HMSO, London.

Davis, J. H. (1957). *Dune Formation and Stabilization by Vegetation and Plantings*, US Beach Erosion Board, Tech. Mem. 101.

Eddington, J. M. and M. A. Eddington (1977). *Ecology and Environmental Planning*, Chapman & Hall, London.

Eddison, J. (1983). The evaluation of barrier beaches between Fairlight and Hythe, *Geogr. J.*, **149**, 39–53.

Findon, R. (1985). Dungeness—ecology and conservation: human pressures. In B. Ferry and S. Waters (eds) *Focus on Nature Conservation*, No. 12, pp. 13–24, NCC, Peterborough.

Fuller, R. M. (1985). Dungeness: an assessment of damage to the shingle ridges and their vegetation. In B. Ferry and S. Waters (eds) *Focus on Nature Conservation*, No. 12, pp. 25–42, NCC, Peterborough.

Gilg, A. W. (1978). *Countryside Planning: the First Three Decades, 1945–1976*, Methuen, London.

Hails, J. R. (1975). Submarine geology, sediment distribution and Quaternary history of Start Bay, Devon: Introduction. *Jl. Geol. Soc. Lond.*, **131**, 1–5.

Hansom, J. D. (1986). Coastal zone management in the United Kingdom. In W. Ritchie, J. C. Stone and A. S. Mather (eds) *Essays for Professor R. E. H. Mellor*, pp. 323–32, University Press, Aberdeen.

Henderson A. (1985). The Lydd ranges. In B. Ferry and S. Waters (eds) *Focus on Nature Conservation*, No. 12, pp. 116–44. NCC, Peterborough.

Herlihy, A. J. (1982). *Coast Protection Survey 1980*, Report, 2 vols, Department of the Environment, London.

Hortig, F. J. (1972). Conservation of mineral resources of the coastal zone. In J. F. P. Brahtz (ed.) *Coastal Zone Management; Multiple Use with Conservation*, pp. 149–89, Wiley, New York.

Ippen, A. T. (1972). Environmental problems and monitoring in coastal waters, *Proceedings, 13th Coastal Engng Conf.*, Vancouver, Vol. 1, pp. 9–32.

Kidson, C. and A. P. Carr (1960). Dune reclamation at Braunton Burrows, Devon, *Chartered Surveyor*, **93**, 298–303.

Kinsey, G. (1981). *Orfordness—Secret Site*, Dalton, Lavenham, Suffolk.

Lewis, J. (1979). *Vulnerability to a Natural Hazard: Geomorphic, Technological and Social Change at Chiswell, Dorset.* Natural Hazard Research Working Paper Series No. 37, University of Colorado, USA.

Moyazaki, M. (1969). Coastal protection works in Japan. *Proceedings, 22nd Inter. Navg. Congress*, Paris. Section II, pp. 133–53.

Parker, D. J. and E. C. Penning-Rowsell (1983). Is shoreline protection worthwhile—approaches to economic variation, pp. 39–57. In *Shoreline Protection*, Proc. Conf. Instn of Civil Engineers, Southampton, 1982. Telford, London.

Penning-Rowsell, E. C. and D. J. Parker (1980). *Chesil Sea Defence Scheme: Benefit Assessment.* Flood Hazard Research Centre, Middlesex Polyechnic, London.

Pilkey, O. H. Jr., W. A. Neal, O. H. Pilkey Sr. and S. R. Riggs (1978). *From Currituck to Calabash.* North Carolina Science and Technology Research Center, USA.

Scott, W. (1985). Dungeness—ecology and conservation: bird populations past and present. In B. Ferry and S. Waters (eds) *Focus on Nature Conservation*, No. 12, pp. 64–93. NCC, Peterborough.

Stahr, E. C. (1974). Defining the National Interest in the coastal zone, *Marine Technol. Soc. J.*, **8**, 10–13.

Stamp, D. (1939). Recent coastal changes in south-eastern England: a discussion. *Geogr. J.*, **93**, 496–503.

Trafford, B. D. and R. J. E. Braybrooks (1983). The background to shoreline protection in Great Britain, pp. 1–8. In *Shoreline Protection*, Proc. Conf. Instn of Civil Engineers, Southampton, 1982. Telford, London.

Williams, A. (1981). *Town and Country Planning Law.* Macdonald & Evans, Plymouth.

Geomorphology in Environmental Planning
Edited by J. M. Hooke



# 12 Coastal Erosion and Flood Control: Changing Institutions, Policies and Research Needs

**EDMUND PENNING-ROWSELL, PAUL THOMPSON and DENNIS PARKER**
*Middlesex Polytechnic, Flood Hazard Research Centre, Queensway, Enfield*

## INTRODUCTION: INGREDIENTS FOR POLICY CONTRIBUTION

If geographers in general—and geomorphologists in particular—are to have an influence on public policy making, several important ingredients are required. Without these ingredients the scientists concerned will lack the authority to advise the relevant agencies of government and private enterprise, and without this credibility they will not make progress.

First and foremost those seeking to have an influence on policy must have technical skills that can be 'sold' successfully in the market place of ideas and technologies. These skills must be publicly recognizable, and useful or necessary inputs to those making policy decisions. They should be supported by a positive record of successful implementation that suggests to those in implementing agencies or government that the advice that they receive will be useful and the best value for money that can be obtained at that time.

Secondly, and also of vital importance, geomorphologists must understand the 'political' circumstances in which their advice is sought. This is an area much neglected in the past, and indeed at present by some researchers who erroneously believe that their technical skills are the prime ingredient and that other considerations are unimportant. Nothing could be further from the truth. To ignore the overall political economy within which policy is made is to ignore the 'laws of motion' by which agencies and their policies evolve in our society. Thus scientists whose recommendations and research can be fitted within the existing institutional and planning framework will have a clear advantage, but this political awareness must not be uncritical. The implications of geomorphological research, for example on the interrelatedness of coastal processes, may point to the need for reforms of this institutional framework and the policies involved. Influence on the currently operating institutional paradigms from the inside will be more effective than remaining aloof from the real world of coastal management decisions.

In reality these two dimensions—technical skills and political understanding—are mutually supportive. Excellence of technical research allows insight by scientists into the inter-agency and intra-agency political situation because the scientist is invited to participate in policy making, either as researcher or as consultant. This 'inside' position will provide the insights that will then contribute to the scientist's understanding of institutions, their problems, and their research and development needs. Therefore, perhaps the prime ingredient for an enhanced policy influence is to develop further the scientist's political antennae, critical perspective, and prescriptive abilities, rather than only their technical skills, which may in any case be more developed than those required for any given situation.

How, then, does this analysis assist our contribution to public policies in our chosen specialist spheres of interest? If the interrelatedness of political understanding and technical skills is accepted, we must develop the former to maximize the utility of the latter. This means enhancing our analysis of institutions: financial structures, the legal framework for policy evolution, the interest groups involved and the orientations of the personnel concerned. Some analysis in this direction has been attempted for the flood hazard and water planning fields (Parker and Penning-Rowsell, 1980; Penning-Rowsell, Parker and Harding, 1986) leading to an enhanced awareness of the issues both in Britain and overeas. However, this chapter is concerned with coastal zone management in its broadest sense, in which new institutions are being encountered and new insights developed.

## COASTAL MANAGEMENT: INSTITUTIONAL COMPLEXITY AND CHANGE

The management of the coast in Britain is acknowledged to be fragmented and ill-coordinated (Gilg, 1978). The primary management task rests with those District and Borough Councils which have sea frontages, but Water Authorities and Port Authorities also have related duties and responsibilities.

### The Separation of Powers

The institutional complexity is compounded by the separation of legislative powers which leads to a confusion as to roles and duties.

Under the Land Drainage Act 1976a, both Water Authorities and Local Authorities have permissive powers for sea defence (against flooding) and under the Coast Protection Act 1949 they are given similar powers for works designed to prevent loss of land through erosion by the sea. The difference between sea defence and erosion control is contentious and a High Court case in 1981 concerning proposed works at Whitstable in Kent to prevent both erosion and sea flooding did little to clarify the situation (Penning-Rowsell, Parker and Harding, 1986, pp. 167–71). The difference is important because different rates of central government grant aid are given under the different Acts and this can radically affect the level of national subsidy given to local coastal management works.

The main powers and resources for coastal management, at least at central government level, are for engineering works and these have dominated the thinking of central and local government in this field. Equally important for integrated coastal management, however, are the land use control powers under the various Town and Country Planning Acts, which can be used to prevent encroachment of urban areas on to those parts of the coast liable to flooding and coastal erosion. The main bodies responsible for implementing this control mechanism are again the local authorities, who will also have interests in promoting recreation on their beaches. However, it is often the case that the different departments within these local authorities have not taken a coordinated view of the development and management of their coastal resources. For example, local land use plans may permit development in areas at risk from erosion, making engineering works justifiable, yet the engineering schemes devised may have adverse impacts on beach access or beach nourishment further down the coast, thereby adversely affecting the potential for tourism in the erosion prone site.

The institutional complexity does not end there. As Carr in Chapter 11 has indicated a further set of interested bodies includes the Nature Conservancy Council and the Countryside Commission who have the national responsibilities for promoting nature conservation and landscape enhancement. There are also the many private organizations and charities concerned with the coasts, the principal landowner of which is the National Trust. However, these organizations are generally not concerned with the day-to-day management of the coastline, and tend to become involved only when these problems are aired, for example at public inquiries, or in the case of the National Trust in acquiring available coastal areas for protection and preservation on an *ad hoc* basis.

## Institutional Change

In 1985, however, a key change in institutional arrangements occurred, which will both add complexity and clarify responsibilities. The driving force behind coastal protection in Britain is central government grant aid to local authorities, which contributes up to 85 per cent of the total costs of the necessary engineering or other works. The task of overseeing coast protection, under the Coast Protection Act 1949, has been transferred to the Ministry of Agriculture, Fisheries and Food (MAFF). The underlying political reasons for this shift are not entirely clear, but part at least of the rationalization of the change was the superior project appraisal methods employed by the Ministry in related fields of investment in sea defence and flood alleviation schemes (for details of these methods see Penning-Rowsell and Chatterton, 1977; and Parker, Green and Thompson, 1987). However, the effect is that the allocation of protective engineering investment by central government is now more unified and rational, with the Ministry now undertaking alone what previously was split between the Ministry (sea defence) and the Department of the Environment (coast protection). At the level of central government this therefore looks like an attempt at unified hazard management.

However, for local government the shift in responsibilities for coastal engineering at central government level has further segmented coastal management responsibilities. Previously part of the grant aiding of their protective works came from the same government Department as was concerned with both land use planning and environmental conservation and enhancement. Now a different Ministry is concerned with coastal works from the one concerned with planning, necessitating different lines of communication and raising the prospect of non-unified or competing policies between the spheres of engineering, recreation and environmental planning.

## Policy Implications

The implications for policy of this change in institutional arrangements will not be completely clear for several years, since—predictably—no explicit policy changes have been announced although several are likely.

First, the Ministry of Agriculture, Fisheries and Food is committed to tightening up the project appraisal methodology for coast protection works—since this appears to have been the basis of their *coup d'état* over the Department of the Environment—in comparison with that required previously by the Department. Little systematic analysis of the costs and benefits of coast protection works was required prior to 1985 (Department of the Environment, 1984), let alone any more extended Environmental Impact Assessment, and much of the previous investment was probably of dubious economic value.

This move to more rigorous assessment will perforce encourage the Ministry to look at the wider aspects of, and returns from, coast protection and management. This is because, as found by the Local Government Operational Research Unit (Mackinder, 1980) few coastal erosion prevention schemes can be justified simply in terms of property losses avoided. If the current level of central government expenditure on coast protection is to continue—and it is in the Ministry's interest to see that it does—then the justification for this expenditure will become more and more concerned with the protection of valuable coastal environments and recreational sites rather than prevention of property damage or loss of developed land through erosion by the sea. The effect will be that the Ministry will be pushed towards playing a larger part in the environmental management of the coast, in the same way that during the last five or ten years it has followed the systematic trend towards integrated river management and away from simplistic river engineering in its work in the flood alleviation field (Newbold, Pursglove and Holmes, 1983).

Secondly, by aligning coast protection, sea defence and river flood alleviation within one Ministry there will inevitably be more scrutiny of the different design or protection standards in the different hazard fields. This may either result in a lowering of some standards at the coast (where protection against sea flooding has often been at standards of the 1 in 150 or even the 1 in 1000 year storm, but many sea walls have been designed to approximately 1 in 70 to 1 in 100 standards), or a raising of standards in the riverine environment. Logic would point to the former, in that high standards cannot usually be justified by any systematic appraisal method owing to lack of extra benefits over more modest schemes. However, such is the domination of the institutions concerned by engineering thinking, and such is their concern with high design standards, that we may well see attempts to raise standards in all areas of 'water hazard' alleviation (i.e. coasts and inland rivers), unless planning and project decisions are more heavily influenced by extended benefit-cost appraisal.

## THREE EXAMPLES OF COASTAL MANAGEMENT COMPLEXITY

Three examples of the complexity of coast protection, sea defence and the institutional arrangements for coastal management arise from involvement by Middlesex Polytechnic in multi-disciplinary research and consultancy teams in highly contentious coastal management projects on different parts of the coastline of southern England. In one case the focus was a major public inquiry, and in the other two there was the threat of serious land losses and flood damage from sudden erosion of an otherwise apparently stable coastline. All have been the subject of years of continuing controversy particularly in the case of Whitstable.

### Chesil Beach, Dorset and Whitstable, Kent

Two studies of potential flooding at the coast have revealed the complexity of appraisals of coast protection and sea defence investments. The first of these concerns the plans of Wessex Water Authority for Chesil Beach (Penning-Rowsell and Parker 1980; Penning-Rowsell, Parker and Harding, 1986) and the second is a long running saga concerning the protection of the Kentish town of Whitstable, and the nearby village of Swalecliffe, from storm surges in the North Sea (Parker and Penning-Rowsell, 1982(a); Parker, Green and Penning-Rowsell, 1983).

In the case of Chesil Beach, flooding was caused by percolation through the beach and overtopping of the beach during storm and surge conditions (Figure 12.1). The key area of data deficiency, other than the usual problem of

**Figure 12.1** Storm surge and extreme wave conditions at Chesil Beach

determining the return periods of extreme events, involved the rates at which water seeped through the beach during normal high spring tides. This percolation caused localized flooding as many as three or four times each year and because of this high frequency many of the economic benefits of investment in sea defence arose from this flood cause (Penning-Rowsell and Parker, 1987).

The consulting engineers employed to determine these seepage rates had great difficulty in perfecting the appropriate experiments to measure this through-beach percolation, not least because of the short time horizon within which such commercial organizations are forced to operate. A proper basic research programme, or database with reviews of similar problems elsewhere (if that were possible), would greatly have aided this task and restrict the problem in scheme appraisal to the simpler one of site specific adaptation of existing models and data.

In the cases of Whitstable and Swalecliffe (Parker and Penning-Rowsell, 1982(a); Parker, Green and Penning-Rowsell, 1983) the most significant single deficiency was determining the joint probabilities of waves and storm surges. In the North Sea the surge pattern is well known and can be predicted using normal extreme probability methods for assessing the heights of high tides and the meteorological conditions likely to give rise to surge events. This is now common practice within the East Coast Storm Tide Warning system used in the operation of the Thames Barrier.

However, when reviewing the combined effects of surges and storms the situation is much more complex. Surges can occur when there are no storms, and *vice versa* (although the largest storms are associated with deep, low pressure cyclones and these also produce surges). There is therefore a correlation between storm surges and wave heights but the correlation is poor, especially for the north-west facing Whitstable beaches. The correlation is better for north-east facing coastline areas in north Kent since surges and storms producing waves then have a common angle of incidence to the coast.

If the *joint* probabilities of high waves and extreme surges are investigated, very low probabilities are the result (i.e. events with return periods well in excess of 1000 years, which is often used as the design standard for coastal defence works). Such extreme events, which reflect the coincidence of large waves plus the high water levels, are important to benefit appraisal and design since they cause major coastal damages. However, this attention to joint extreme events should not be at the cost of neglecting the single events, since the latter events may cause less damage than combined events but occur more frequently and so may be of greater significance economically and environmentally. Nevertheless, if we combine the requirement for joint probabilities with the need to allow for a rising sea level, which results in the return period of a given event falling over time, it is possible to appreciate the true complexity of this analysis.

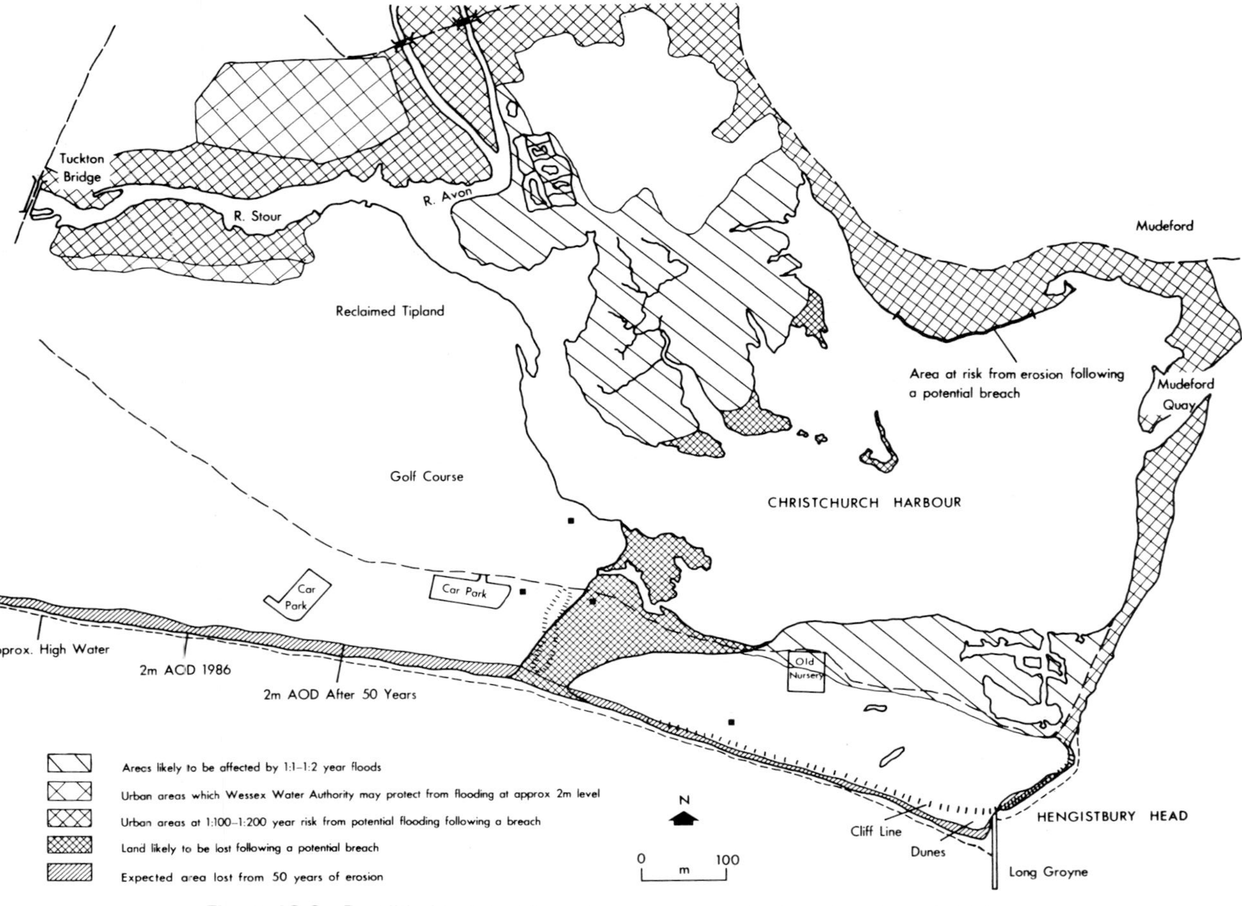


**Figure 12.2** Possible impacts of erosion and a potential breach at Double Dykes

**The Possible Breaching of Christchurch Harbour at Hengistbury Head/Double Dykes, Dorset**

The coastline to the west of Hengistbury Head in Dorset is eroding rapidly. It is possible that this erosion is exacerbated by protection of the beaches further to the west, at Bournemouth, which have been stabilized by a system of groynes which retain the sand to preserve the beach for recreation. The area being eroded is to the east of the last of these groynes, and erosion of the cliffline has been noted to be in excess of 1.0 m per annum.

Whatever the cause of the erosion there is a risk of the sea breaking through the low cliffs just to the west of Hengistbury Head, which are made of soft unconsolidated alluvial deposits. Waves breaking over these cliffs during storms at high spring tides would produce overland flow northwards into Christchurch Harbour. If large enough this flow would erode a gully through the neck of land between the Head and the 'mainland'. Bournemouth Borough Council plans to install further groynes coupled with beach nourishment both at Double Dykes, to prevent the breach from occurring, and east of the Long Groyne to prevent erosion of the Head itself (Figure 12.2). The scheme, if approved, would be supported by government grant aid from the Ministry of Agriculture, Fisheries and Food.

Without protection of the coast at this point a number of impacts are anticipated. If the sea breaks through the existing cliffline, to leave the Head as an island, the effect on Christchurch Harbour would be profound. Firstly, large areas of the foreshore of the harbour would be liable to flooding. Tides inside the harbour are now moderate in their amplitude owing to the very restricted harbour entrance (Figure 12.2). With the larger entrance created by the breach the amplitude of the tides would approximate to that experienced in the adjacent sea, yielding much higher storm tides than are currently experienced in the harbour. The exact areas affected by this flooding would depend upon whether the River Stour flood alleviation scheme planning by Wessex Water Authority is completed by the time the breach occurred, since this scheme would afford Christchurch some protection against the more frequent sea floods. In any event, however, the areas prone to flooding would be substantial and the damage would be serious (Thompson and Parker, 1986).

Secondly, areas of the harbour's northern foreshore would be eroded by the waves entering the harbour through the breach (Figure 12.2). Because the harbour entrance would then be south-west facing it would catch the full force of waves from the English Channel and the Atlantic Ocean.

The full impacts of this beach and cliff erosion, and the threat of breaching, have been investigated and described in a substantial research report to Bournemouth Borough Council (Thompson and Parker, 1986). As indicated in Table 12.1 these impacts are both many and interrelated. The consequent flooding would cause serious damage in Mudeford, and render

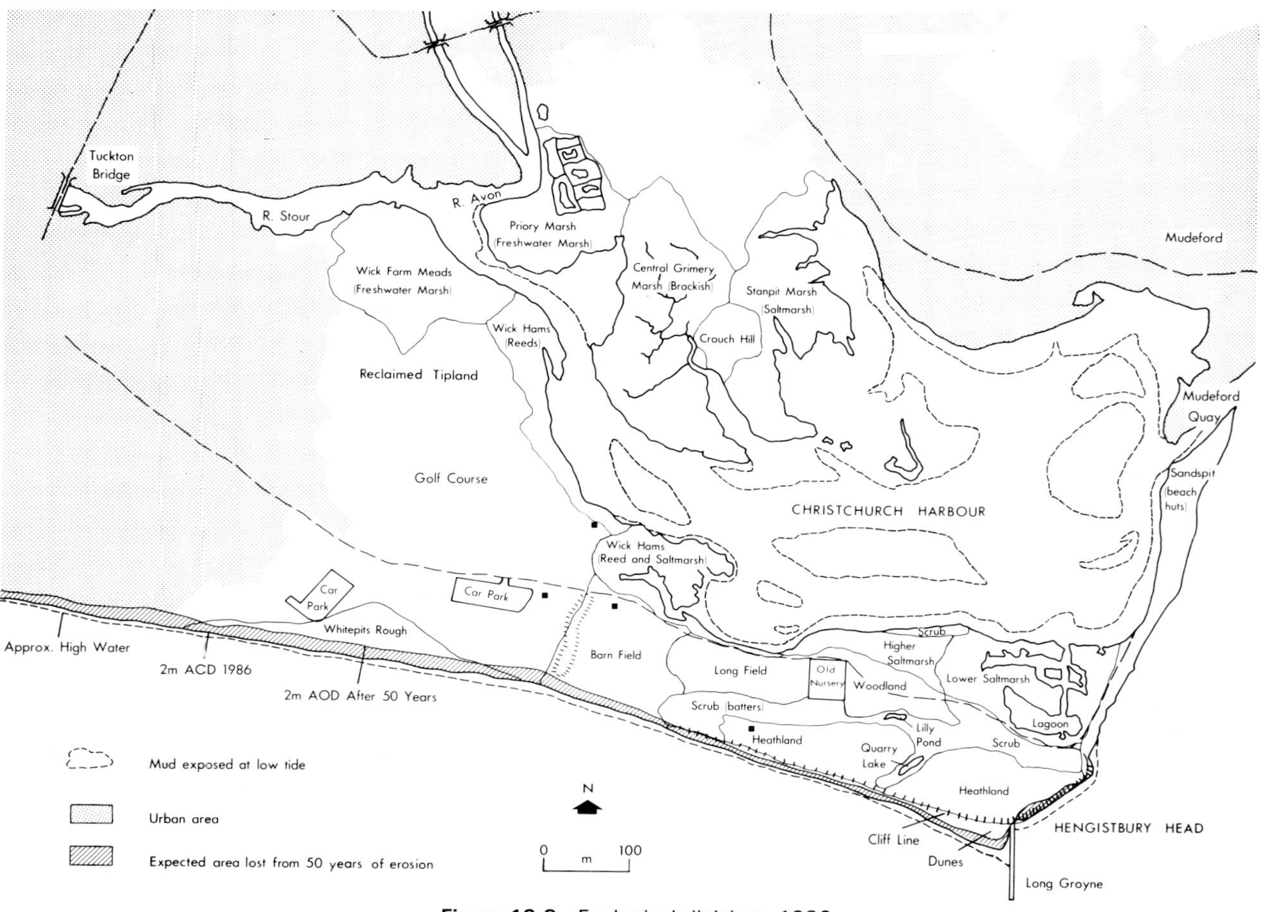


**Figure 12.3** Ecological divisions 1986

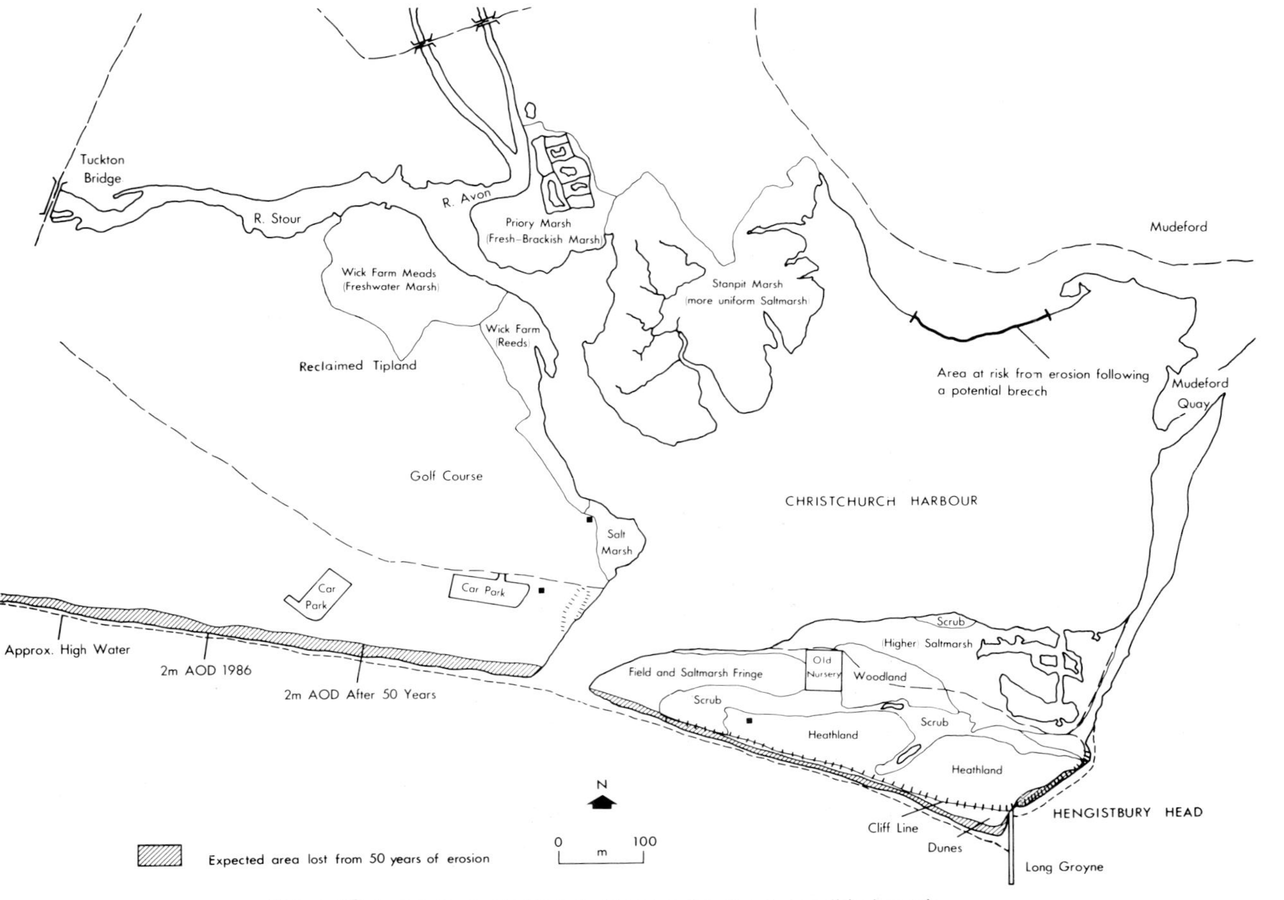


**Figure 12.4** Likely ecological divisions following a possible breach

**Table 12.1** Hengistbury Head, Dorset: Impacts averted by coast protection works

| Impact | Immediate | Subsequent |
|---|---|---|
| 1. Potential flood losses<br>direct<br>indirect | | ★ |
| 2. Property loss—unusable owing to frequent flooding | | ★ |
| 3. 'Social impacts' such as possible stress and ill-health of affected community | | ★ |
| 4. Land and property loss from erosion | ★ | |
| 5. Loss of British Telecom cables | ★ | |
| 6. Ecological impacts from flooding | | ★ |
| 7. Ecological impacts from erosion | ★ | |
| 8. Archaeological impacts from erosion | ★ | |
| 9. Loss of recreation potential of Head | | ★ |
| 10. Impact on use of Christchurch Harbour (commercial and recreational) | | ★ |
| 11. Geological impacts of project itself (Tertiary exposures east of Long Groyne) | ● | |

*Notes:* ● = negative impact of protection.
Immediate = direct impact of erosion.
Subsequent = consequence of a breach.

certain property there useless, since it would be flooded virtually every year. In addition, there would be major changes to the plant and animal communities inside the harbour area, where currently there is a complex pattern of freshwater and saltwater habitats, including extensive salt marshes (Figures 12.3 and 12.4). The result would be a diminution of ecological variety and quality and a total loss of certain habitats, principally the lower saltmarsh at Hengistbury and the maritime grassland at Double Dykes.

In addition the recreation potential of the area would be drastically reduced after the breach had occurred, and this would be a serious loss given that annually there are approximately 740 000 visitor/days spent on the Head. The attractions are: walking, general leisure pursuits, the natural environment and the dramatic coastal scenery. The recreational use of the harbour would also be restricted, with reduced facilities for windsurfing and dinghy sailing owing to the loss of sheltered water space. Commercial fishing would have its costs increased although fish stocks would probably be unaffected.

The erosion of the coastline would cut British Telecom cables to the Channel Islands, although these are now virtually at the end of their useful lives. More serious would be the loss of archaeological remains. Hengistbury Head is an area of outstanding archaeological importance, comprising a complex mosaic of remains covering the major periods from the Upper Palaeolithic to the late Iron Age; it is also of interest for nineteenth-century industrial archaeology (Figure 12.5).

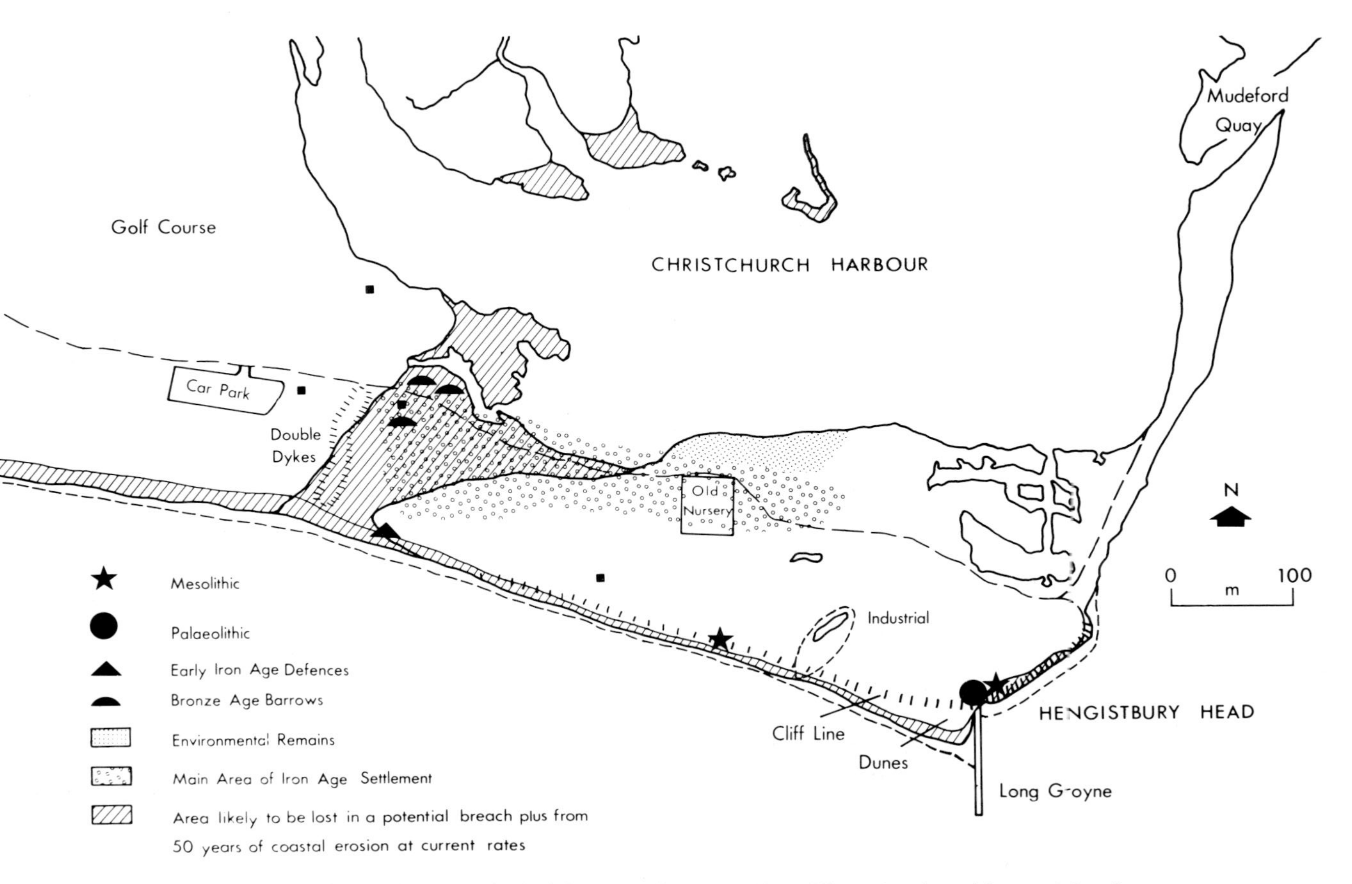


**Figure 12.5** Principal archaeological features threatened by cliff erosion breaching and flooding

The assessment of the benefits of the coast protection schemes proposed for Hengistbury Head shows that investing up to £3.7 m to prevent a breach is worthwhile even on the basis of only the benefits in the form of direct flood damages, property loss, and other evaluated losses such as recreational impacts (Thompson and Parker, 1986). This assessment is based on estimated cliff erosion rates and a secular sea-level rise of 1.5 mm per annum over the proposed fifty-year design life of a coast protection scheme. This result is obtained despite the possibility that a breach might not occur during the fifty-year period considered in the evaluation; changes in the rate of erosion and sea-level rise profoundly affect the risk of a breach over time and thereby the present value of the discounted benefits (Figure 12.6).

The 'extended cost–benefit analysis' undertaken gives a ratio of benefits to costs of between 1.55:1 and 0.88:1 (but the latter excludes all recreational benefits), excluding the net gains from preserving the environmental and archaeological values of the area as a whole. Within this range the best conservative estimate of the benefit–cost ratio is either 1.12:1 or 1.07:1 (again excluding the environmental gains) depending upon whether or not the proposed River Stour flood alleviation scheme is implemented. However, these extended

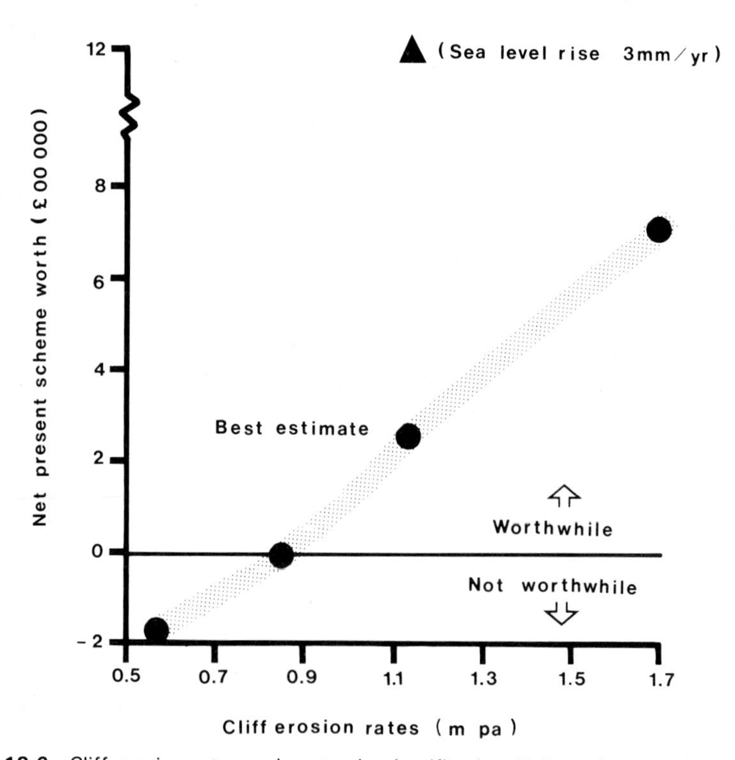


Figure 12.6 Cliff erosion rates and protection justification (1.5 mm/p.a. sea-level rise)

benefit–cost results are again very sensitive to the assumptions about cliff erosion rates and other data supplied by Hydraulics Research Ltd for the probability of the breach and the extent of flooding.

Many lessons emerge from the Hengistbury appraisal. These concern the methodology of appraising coast protection scheme in general, as well as the interrelatedness of the likely impacts, which will vary from location to location. As far as the methodology is concerned, the study emphasizes that losses from coastal erosion are essentially '*one-off*', in the sense that once erosion has removed a building (or an archaeological remain) from the shoreline it cannot be recovered. In contrast, flooding is a repeating event; coastal erosion is *irreversible* in that damage occurs once and will not be repeated.

Secondly, damage or losses are merely *delayed* by coast protection works, rather than prevented, and they are *sequential* as erosion moves inland. The pattern of damages is dependent on the pattern or sequence of erosion lines, the pattern of land use within those areas, and of course the geomorphology of the area being eroded. Even property at some distance from the shoreline benefits from coast protection, because that protection delays its eventual loss through erosion. However, depending on the pace of erosion, and the discount factor used, it is possible to ignore those distant effects.

Two other factors are important. First, the valuation of property at risk from erosion is problematic because that erosion threat may depress market prices for the property in question. Which value to take in any project appraisal is controversial and, to date, the depressed prices have not been used. Secondly, given that erosion is a sequential process and that many properties are set back geographically from the point of erosion, it may be sensible, in economic terms, to delay investment until nearer the time when losses are imminent; certainly the benefit–cost ratio for such a project will rise as it is delayed and, over time, erosion proceeds and valuable property gets closer to the point at which it will suffer damage or loss (Thompson *et al.*, 1987).

A final set of lessons concerns the relationship between the appraisal results and the erosion rates estimated for the coastline at Double Dykes. Estimates of such erosion rates are necessarily crude. In the Hengistbury appraisal these rates were assessed based on historical maps, which show the evolution of the shoreline since the nineteenth century, but this assessment is bound to be rudimentary.

The precise effect of the breach on the harbour is also important to the assessment, and yet again is based on rudimentary estimates rather than systematic modelling. Only more detailed monitoring of the beach at Double Dykes, and more sophisticated assessments of the geomorphology of the harbour, would allow improvements in this aspect of the assessment.

In addition, the rate of secular sea-level rise profoundly affects the appraisal results. A best estimate of a rise of 1.5 mm per annum was used for the appraisal, but if that assumed rate is doubled the worthwhileness of the scheme rises dramatically (Figure 12.6). An important additional difficulty with this particular Hengistbury appraisal, but one that may have more general implications, is that it is relatively straightforward to assess the impact of one variable on the complex situation (e.g. beach/cliff erosion rates). However, assessing the combined effects of erosion and overtopping of the cliff with gully-creating spray and overland flow is not easy. Most difficult is assessing the effects of storm events which are less than those which have the magnitude necessary to cause a breach but which will probably have an effect on increasing the likelihood of breaching by weakening the coastline through minor gullying.

## A CHANGING RESEARCH AGENDA

Parallel to the changing institutional arrangements are coming new directions in research and development work concerned with coastal protection, reflecting in part the concern of the Ministry of Agriculture, Fisheries and Food to enhance understanding in this complex area, and to make a demonstrable impact on this policy field. There is no doubt that the land/sea interface contains natural systems of extreme complexity and far less is known about their processes and effects than is the case with inland areas.

Hence there is a need perceived in many quarters for more fundamental and applied research in this area from a wide range of disciplines. Two examples of this new reseach agenda are outlined below, plus an analysis of necessary advances in the physical science.

### The CIRIA Review — Engineering Design

The Construction Industry Research and Information Association is coordinating a major review of existing sea-wall design procedures and standards, sponsored by the Ministry of Agriculture, Fisheries and Food at a cost of over £100 000. The objective is to select from the many techniques and approaches, mainly developed by Hydraulics Research Ltd and engineering consultants, the best methods that can be used in designing coast protection works (Lewis and Duvivier *et al.* 1986).

The approach adopted has been to survey all known active Coast Protection Authorities (largely District Councils) to determine the problems that they have with respect to sea-wall design. A parallel exercise by consulting engineers Lewis and Duvivier is reviewing the international literature on coast protection, to determine best practices elsewhere. The result will be an analysis of the shortfalls in design practice as used in Britain's coastal authorities, and recommendations for both further research and the spreading of known techniques to a wider audience.

## The Coast Protection Research Project (COPRES) – Appraisal Techniques

Existing research is deficient in respect both to data on coastal erosion and flooding, and the methods for investment appraisal. The COPRES project now underway at Middlesex Polytechnic is intended to ensure that coast protection authorities have better advice, data and methods on which to base their new policies. The rationale of the project is that the coast of Britain suffers from both erosion and periodic flooding, and that the two are interrelated.

The context of this research project is broad. Flooding from the sea can have a devastating effect, as was shown by the disastrous East coast floods in 1953. The defences raised after this event, which cost over 300 lives, are now nearing the end of their design lives. Further damaging floods could follow if these defences were not replaced and if the rise in sea levels in relation to the land continues (Penning-Rowsell, Parker and Harding, 1986).

In certain locations such as the Holderness coastline in North Humberside, whole communities are at risk from erosion of the land by the sea. Elsewhere tourist trade of coastal towns is threatened by erosion removing their beaches. In a crowded island such as Britain land is valuable and this type of loss to the sea can be serious. However, the coast is also an area of immense ecological value and natural beauty. For example, Sites of Special Scientific Interest can be threatened by erosion by the sea, but also by unwise engineering or other works to counter flooding and erosion.

Central Government, Coast Protection Authorities, and other agencies such as the National Trust have therefore to develop appropriate policies to protect and manage coastal areas from nature's own force of destruction, as well as from the intense recreational pressure at Britain's crowded coastline. The resource costs and impact on the environment of such policies and works must be balanced against the benefits of retaining the valuable coastal land resources and preventing unnecessary flooding. The COPRES project is designed to provide new methods, data and information for evaluating the gains to society from coastal works (Thompson *et al.*, 1987).

The research on coastal *flooding* is examining the flooding experienced around Britain's coastline and the consequential damage caused. Interviews are being held on a controlled basis with those suffering damage and those living or working in areas liable to flooding to assess damage to property, damage to protection structures such as sea walls, loss of trade from flooding-induced disruption, 'intangible' damages from sea flooding (flood victims' stress), and possible loss of life. For the investigation of *erosion* caused by the sea we are studying the loss of property, communications and land, together with the erosion's environmental and recreational effects.

In each case the objective is to derive data and information from case studies of erosion and flooding already experienced. This will be used to generate a database of general applicability to evaluate future investment in coast protection and sea defence elsewhere.

### Geomorphological Research Needs

Currently the integration of geomorphological research into coastal processes with coast protection engineering proposals is weak and inadequate, and here geomorphologists have a major opportunity to apply their expertise to coastal management (Ricketts, 1986).

From the point of view of project and policy appraisal in the coastal field, the key research topics for the physical sciences should be to alleviate some of the data deficiencies identified in the three examples above. Research needs to include modelling of erosion rates for different geological substrata and varying wave and storm conditions. More knowledge is also needed on the stability and slippage characteristics of different rock types.

More importantly, modelling of coastal processes at a macro level is vital—for example on the extent to which erosion feeds beaches down-coast—and on the extent to which protection works alter erosion and deposition patterns both at and away from the protected section of coast. New information on these so-called 'externalities' is vital if the engineering works implemented around our coasts are to be planned successfully for the overall good of society.

## CONCLUSIONS

The implications of the three case study examples discussed above are that project appraisal at the coast, designed to achieve better coastal management, is not a simple exercise dealing with a single discipline or even a common approach. The pursuit of economic efficiency cannot be the only objective (Parker and Penning-Rowsell, 1982(b)) when so many of the effects of coastal erosion and flooding have wider implications than property damage, as is the case for many of the effects of the coastal management plans and schemes themselves. Many policies and appraisal techniques require attention, so that coastal management is more sympathetic to the needs of those concerned to protect the environment, commensurate with preventing both wasteful erosion of the land and flooding from the sea which can cost lives and many millions of pounds of damage.

Coupled with this urgent requirement for more sophisticated analysis of coastal management needs comes the change in institutional structures, focusing on the changed role of the Ministry of Agriculture, Fisheries and Food. This change needs to be complemented by a change in the Ministry's emphasis and perhaps in personnel, to reflect the wider environmental role that coastal management must bring. This comprises a broadening of horizons away from simplistic benefit–cost analysis and engineering design, towards a more extended environmental and economic appraisal of investment and management needs.

The potential for geomorphologists to have a more active role in research and policy making is substantial. However, this must be based on a high level of technical skills so that they can contribute to ameliorating just those types of deficiencies in techniques and data that the examples above so clearly reveal. In this respect these geomorphologists could be equally as well placed as the other scientists and engineers who currently dominate the scene within organizations such as Hydraulics Research Ltd and the expert consulting engineers. Many coastal management and protection projects require just the kind of broad appreciation of geographical context with which to complement the detailed specialist skills that the geomorphologist can bring, rather than the narrower and solely technical base that often is the characteristic of engineers and hydraulic specialists.

However, this new influence on policy will not materialize unless geomorphologists devote more attention to the political interests within both the research organizations and the government agencies in the coastal management sphere. Developing a perceptive understanding of those interests, and the processes linking the institutions involved, is just as important to making progress in influencing policy as having the technical skills so badly needed to improve our predictions of the physical evolution of the land/sea interface.

## REFERENCES

Department of the Environment (1984). *Notes of Guidance in the Preparation of CBA for Works Proposed under the Coast Protection Act 1949*, Department of the Environment, London.

Gilg, A. W. (1978). *Countryside Planning: the First Three Decades, 1945–1976*, Methuen, London.

Lewis and Duvivier, and Posford, Pavry and Partners (1986). *Sea Wall Design Guidelines: Inception Report*, Construction Industry Research and Information Association, London.

Mackinder, I. H. (1980). *The Economics of Coast Protection*. Report C284, Local Government Operational Research Unit, Reading.

Newbold, C., J. Pursglove and N. Holmes (1983). *Nature Conservation and River Engineering*, Nature Conservancy Council, London.

Parker, D. J., C. H. Green and E. C. Penning-Rowsell (1983). *Swalecliffe Coast Protection Proposals: Evaluation of Potential Benefits*, Middlesex Polytechnic Flood Hazard Research Centre, London.

Parker, D. J., C. H. Green and P. M. Thompson (1987). *Urban Flood Protection Benefits: a Project Appraisal Guide*, Gower Technical Press, Aldershot.

Parker, D. J. and E. C. Penning-Rowsell (1980). *Water Planning in Britain*. Allen & Unwin, London.

Parker, D. J. and E. C. Penning-Rowsell (1982a). Flood risk in the urban environment. In D. T. Herbert and R. J. Johnson (eds) *Geography in the Urban Environment*, John Wiley, Chichester.

Parker, D. J. and E. C. Penning-Rowsell (1982b). Is shoreline protection worthwhile?—approaches to economic valuation. In *Shoreline Protection*; proceedings of a conference of the Institution of Civil Engineers, Southampton, pp. 39–57, Thomas Telford, London.

Penning-Rowsell, E. C. and J. B. Chatterton (1977). *The Benefits of Flood Alleviation: a Manual of Assessment Techniques*, Gower, Aldershot.

Penning-Rowsell, E. C. and D. J. Parker (1980). *Chesil Sea Defence Scheme: Benefit Assessment*, Middlesex Polytechnic Flood Hazard Research Centre, London.

Penning-Rowsell, E. C., D. J. Parker and D. M. Harding (1986). *Floods and Drainage: British Policies for Hazard Reduction, Agricultural Improvement and Wetland Conservation*, Allen & Unwin, London.

Penning-Rowsell, E. C. and D. J. Parker (1987). The indirect effects of floods and benefits of flood alleviation: evaluating the Chesil Sea Defence Scheme, *Applied Geography*, **7**, 263–88.

Ricketts, P. J. (1986). National policy and management responses to the hazard of coastal erosion in Britain and the United States, *Applied Geography*, **6**, 197–221.

Thompson, P. M. and D. J. Parker (1986). *Hengistbury Head Coast Protection Proposals: Assessment of Potential Benefits and Costs*. Two vols (Summary Report and Supplementary Report), Middlesex Polytechnic Flood Hazard Research Centre, London.

Thompson, P. M., E. C. Penning-Rowsell, D. J. Parker and M. I. Hill (1987). *Interim Guidelines for the Economic Evaluation of Coast Protection and Sea Defence Schemes*. Middlesex Polytechnic Flood Hazard Research Centre, London.

Geomorphology in Environmental Planning
Edited by J. M. Hooke



# 13 Coastal Erosion: Protection and Planning in Relation to Public Policies—a Case Study from Downderry, South-east Cornwall

**PETER SIMS and LES TERNAN**
*Department of Geographical Sciences, Plymouth Polytechnic*

## INTRODUCTION

The boundary between land and sea is often a zone of conflict not only as a result of the varying levels of energy input by oceanic and atmospheric forces but also on account of human use of the coastal zone. Frequently, and quite unreasonably, we expect the boundary between land and sea to remain static. Such a view is aided by the often imperceptible nature of sediment movement and subsequent coastal changes. However, in active geomorphological zones where waves regularly undercut cliffs and trim beaches or where high pore-water pressures and saturation by groundwater facilitate cliff collapse, unsuspecting coastal dwellers can be jolted into action by the sudden realization that a receding shoreline is threatening their property or livelihoods (Clark, Ricketts and Small, 1976).

In particular, erosion becomes perceived as a natural hazard when residential or commercial developments are threatened by sudden cliff failures or when there are large-scale removals of beach materials by storm waves and flooding associated with a tidal surge. It is at this stage that erosion assumes a social and economic significance and becomes associated with both local and national government strategies for dealing with such hazards. In turn both collective and individual responses have become influenced by public policy (Ricketts, 1986).

The aim of this chapter is to show through a detailed examination of the 1 km coastal sector between Seaton and Downderry in south-east Cornwall (Figure 13.1) how geomorphological factors generating active coastal recession have interrelated with human responses and public policies, and to recommend how geomorphologists can aid the design and implementation of coastal management policies. The first section of the chapter deals with the physical setting of the case study, rates of cliff recession, active geomorphological processes, and the geotechnical properties of the cliff material. There is also a brief consideration

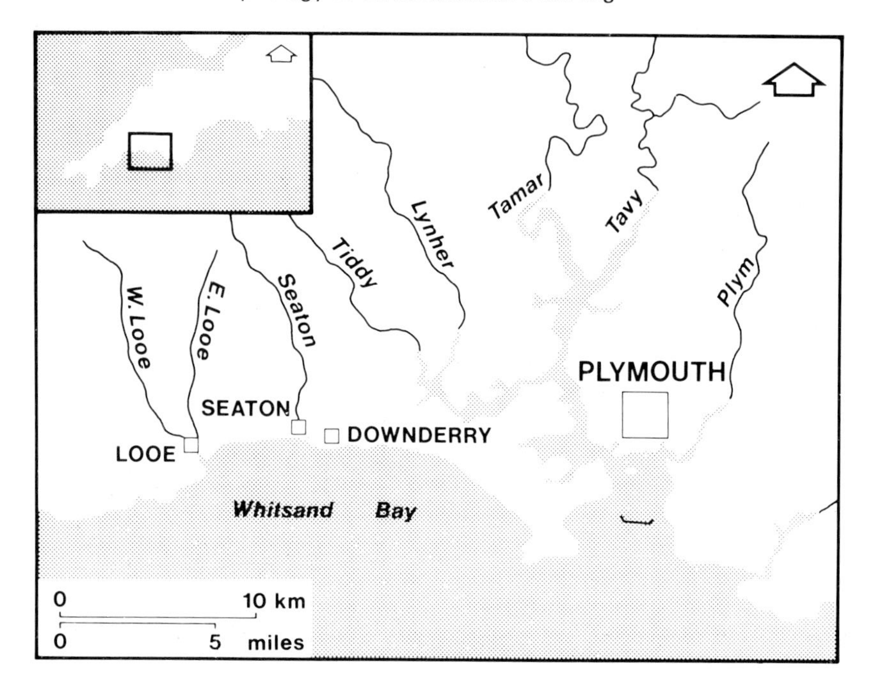


**Figure 13.1** Location map

of the influences of human activity. The second part of the chapter deals with the history of coastal erosion, protection and planning and indicates how these are managed in terms of public policies.

## PHYSICAL BASIS

The 1 km section of the south Cornwall coast between the villages of Seaton in the west and Downderry in the east (Figure 13.1) largely consists of a solifluction terrace overlying a wave-cut rock platform. This platform truncates steeply dipping lower Devonian slates, and occurs at or just above the present day high-water mark. Consequently, for much of this coastal strip the cliffs consist entirely of the unconsolidated sediments forming the terrace. The terrace attains a maximum width of 130 m at Downderry (Figure 13.2) with a cliff height of 6 to 8 m. The cliff height progressively increases westwards to 20 to 25 m with a corresponding decrease in terrace width. The final 250 m of cliffline to the east of Seaton, now protected by a County Council sea wall (Figure 13.2) is in slate bedrock, the solifluction terrace having been entirely removed by marine action.

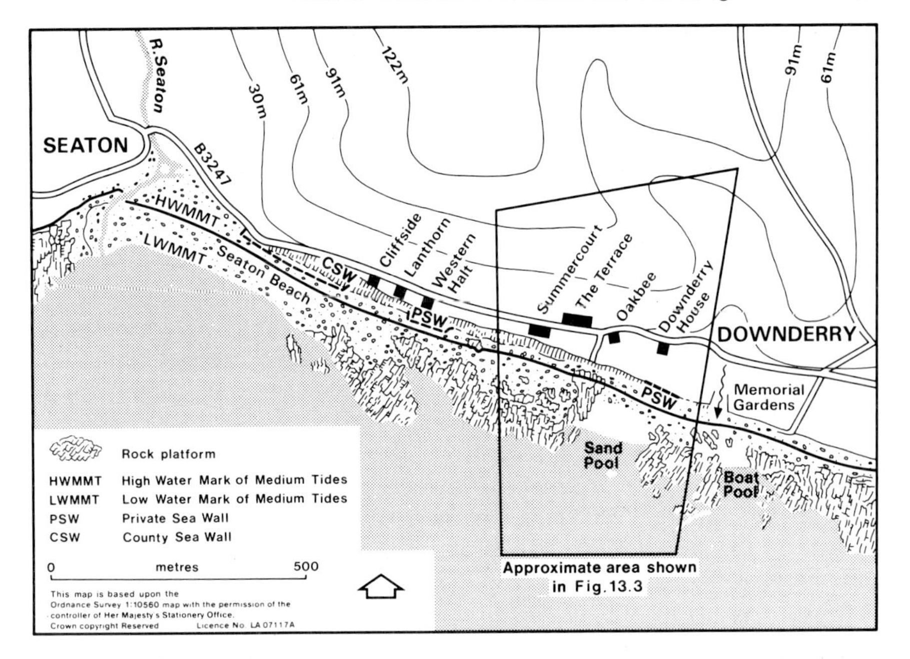


**Figure 13.2** Coastal sector between Seaton and Downderry

## Long-Term Rates of Cliff Retreat

Long-term rates of cliff retreat were obtained by comparing successive positions of the cliff-top with respect to fixed points inland using the 1845 tithe map and the 1966 revised edition of the OS 1:2500 map. Although the technique may be subject to some inaccuracy arising from lack of precision in the tithe map, these data are in accord with more recent field observations of cliff recession. Rates of recession over the 121-year period appear to be greatest in the western part of the sector with 4.5 m (3.7 cm/year) below Oakbee (Figure 13.2), 9.5 and 9.8 m (7.9, 8.1 cm/year) at Summercourt and Western Halt to 13.6 m (11.2 cm/year) at Cliffside at the western limit of the terrace. These data are comparable with an average of 4.8 cm/year and a maximum of 11.9 cm/year determined by Lemon and Blizard (1974) for an 84-year period. A public enquiry in 1959–60 had cited a much faster rate of retreat of 30.5 m (100 ft) of erosion in 106 years (28.7 cm/year) but the basis of this figure is unknown. A gap of 2.9 m between the present day cliff line and the remnants of concrete steps constructed in 1960 as part of a path from the beach to the cliff-top provides a figure of 10.7 cm/year for more recent years.

Such a rate of recession can appear more dramatic as much of this occurs in single failures, with collapses of varying magnitude separated by periods of inactivity, rather than continuous progressive retreat. Lemon and Blizard (1981) in fact reported that the general appearance of the cliff and the indicators of instability had not altered markedly since their 1974 report. Furthermore the toe of a 1974 rotational slip had only been subject to minor trimming in the intervening seven years. The evidence therefore is of a relatively slow long-term rate of cliff-line retreat.

### Form and Causes of Cliff Retreat

Two principal forms of slope failure and cliff recession have occurred along this strip of coast. On the higher cliffs at the western end a complex rotational failure took place in 1974 leading to an abrupt recession of the cliff-top by approximately 11 m and the gardens of a cliff-top dwelling (Cliffside) dropping by 20 m. No further rotational movements have occurred since. East of Western Halt (Figure 13.2) cliff recession is in the form of shallow planar landslips with wedge-shaped failures occurring along near vertical fissures, usually less than 2 m depth from the cliff face. Tension cracks commonly develop behind the uppermost 2–3 m of the cliff.

Cliffs may only be supported to certain critical heights depending on the relationships between the shear strength of the cliff materials and the cliff angle. A general formula linking this critical height and cliff angle has been described by Carson and Kirkby (1972) based on Culmann (1866). Using data on the geotechnical properties of the cliff material it is evident that the cliffs are substantially higher than the critical failure height. Consequently continual removal of slipped material from the base of the slope by marine activity will ensure continuing cliff failure. These failures predominate in late winter and early spring when groundwater levels are at a maximum and give rise to seepages from above an impermeable silt-rich horizon at the base of the solifluction deposits. Thus the main natural cause of these various failures appears to be a combination of marine activity aided by the characteristics of the head materials and by adverse groundwater conditions.

The deposits have suffered continued marine erosion from two main causes. Firstly there has been a progressive rise in sea levels over the last 15 000 years, which according to Thomas (1985) is still occurring at a rate of 2 mm per year in South-West England. Secondly, superimposed upon this are the influences of surges and modern storm wave activity. Evidence derived from Ordnance Survey maps at 1:1250 and 1:10 560 scales of 1880, 1907, 1952 and 1964 indicates that the width of the intertidal zone between mean high water ordinary tides and mean low water ordinary tides has been reduced by approximately 50 m, equivalent to a rate of 0.7 m per year. Much of this is due to a rise in the mean low water ordinary tide line. Furthermore the 1966 revised edition of 1:25 000 OS map shows increased areas of exposed rock as compared with the 1907 second edition, particularly in the vicinity of Sand Pool and Boat Pool (Figures 13.2

**Figure 13.3** Part of the coastal terrace between Seaton and Downderry.

Key: (1) Coastal Hills of Devonian slate. (2) Solifluction terrace, with westward increase in cliff height and decrease in terrace width. (3) B3247 coast road approximately on break of slope at back of solifluction terrace. (4) Exposed wave-cut platform. (5) Modern beach sediments. (6) High water mark ordinary tides. (7) 'Sand Pool'. (8) Private seawall now partly collapsed below Summer Court. (9) Private seawall below Downderry House. (10) Remains of 1960 cliff path steps. (11) Site of new properties on seaward side of coast road in front of The Terrace (buildings now in occupation)

and 13.3). Although this map evidence suggests that depletion of beach sediment occurred prior to 1966, plotting of areas of exposed rock platform from aerial photographs taken in 1964, 1968 and 1981 has revealed an overall decrease in the area of exposed rock with increased sand deposition in the nearshore zone.

Although it is not possible to ascertain from these 'snapshots' whether such changes are indicative of other than short-term variations in the nearshore circulation of sediments and longshore drifting, the cessation of sand and gravel abstraction from Seaton beach in 1960 following a local public enquiry could account for the increased sand deposition since that time. However, the construction of a small sea wall to the west of Seaton and the major coastal protection works further west at Looe might have been expected to cut off some local sediment supply brought to Seaton–Downderry by an easterly longshore drift. In addition to the changes in natural beach protection afforded by sediments, the occurrence of rock buttresses on the wave-cut platform appears to be important. Although difficult to isolate from the effects of small private coastal protection structures these buttresses seem to coincide with areas of more stable cliff as indicated by a more mature vegetation cover (Figure 13.3). They therefore appear to be significant in reducing wave energy and subsequent erosion.

### Human Activity

Although the natural environmental conditions along this section of coast indicate a high vulnerability to coastal recession processes, it is undoubtedly true that human activity has increased the hazard. Construction of cliff-top dwellings will have increased the shear stresses on the cliff materials. More significantly, however, the practice of channelling stormwater from roofs and paths into soakaways in the head materials will have greatly increased the landslide hazard. This view is supported by Consulting Engineers' reports on the Seaton–Downderry area which have consistently pointed out that high groundwater levels in the cliff deposits can seriously reduce the *in situ* shear strength of the materials (Lemon and Blizard, 1974, 1981). Furthermore, the major failure surface of the 1974 landslide intersected the terrace surface within a few metres of the soakaway of the main property affected. In view of this it seems somewhat surprising that the recent infill property development along this cliff top continues to dispose of storm runoff into soakaways.

## ASPECTS OF LOCAL COASTLINE MANAGEMENT

The period since the early 1950s has seen a history of cliff recession alongside continued local pressure for coast protection. Despite the major landslide in 1974 which led to the demolition of a cliff-top property no action was taken until 1981. Following the appearance of cracks in the B3247 highway a 270-metre sea wall was constructed by Cornwall County Council to protect the road. Although it was planned that this sea wall would be extended by Caradon District Council to protect the cliff-top residential properties this has not so far occurred. The concern expressed both at a public enquiry in 1960 and more recently in 1982–3 was the difficulty of reconciling the high capital costs of a sea wall with the actual value of the properties. Throughout, however, village representatives had indicated that a less grandiose and presumably less costly scheme might be acceptable. Only since 1985 when the responsibility for funding coastal protection schemes passed to the Ministry of Agriculture, Fisheries and Food have low-cost options been seriously considered, and investigated by Sims and Ternan (1987). Paralleling the debate concerning cliff recession and coast

**Table 13.1** Chronology of recent coastal protection problems, cliff failures and residential developments

| | |
|---|---|
| Early 1950s | County Council plan to re-route B3247 coast road dropped in favour of coast protection option. |
| 1959–60 | Public enquiry held by St Germans RDC concludes<br>1. 100 ft of erosion had occurred in previous 106 years and was ongoing.<br>2. Erosion due to 'weather' at western end; combination of sea and weather at eastern end.<br>3. Sand and gravel abstraction was of no consequence.<br>4. No justification for £50 000 expenditure in view of the value of properties. |
| 1971 | St Germans RDC carry out gabion basket protection of 60 m of cliff below Memorial Gardens. |
| 1973 | Planning permission granted for Cliffside extension at western end of terrace. |
| 1974 | Cliff failure below Cliffside exposes storm soakaways and renders property uninhabitable. Subsequently demolished. |
| 1974 | Consulting engineers recommend 5 stage coast protection scheme, 755 m sea wall (£829 000 estimate) and 500 m gabion basket (£183 000). |
| 1977<br>1979–80<br>1982–3 | Planning applications result in construction of two bungalows on cliff-top despite local representations concerning cliff recession and need for stormwater drainage. |
| 1977–79 | Continuing discussions on protection schemes. Residents Association write to MP recommending 'a less grandiose scheme'. |
| 1981 | Cracks appear in B3247. |
| 1981–3 | Construction of County Council sea wall to protect B3247 at three times 1974 estimate. |
| 1982–3 | Continued pressure on District Council to extend sea wall to protect residential property. |
| 1983 | Further landslides. |
| 1983–4 | Stage 1 (155 m) costs now estimated at £576 000 to protect properties valued at £81 250. |
| 1985 | Funding responsibility shifted from DoE to MAFF. Caradon DC lose £40 000 costs related to production of sea-wall plans for DoE; MAFF considers a low-cost scheme more appropriate. |
| 1985 | A further planning application refused on grounds of plot size and land instability. |
| 1987 | Caradon DC commission report on low-cost coast protection options. |

protection has been the granting by the District Council of planning permissions which have resulted in the construction of several new bungalows on the cliff top. The view of the planning officer was that the principal responsibility for protecting private property lay with the owner, and that planning permission could not be refused on the grounds of the hazard to property presented by cliff recession.

## PUBLIC POLICIES AND COASTAL EROSION AND PROTECTION

Coastal erosion presents many problems for potential management in that human interventions often exacerbate erosional activity. This tendency is particularly noticeable in connection with sediment starvation where the construction of protective sea walls has the effect of cutting off the natural sediment supply and thus natural beach replenishment. Often particular combinations of geomorphological processes can lead to circumstances which produce short-term intensifications of erosion without disrupting the long-term dynamic equilibrium. Demands for coastal protection may be induced by such short-term variations resulting in schemes which often have geomorphological implications. For example, the removal of sediment from a beach by concentrated storm wave activity can induce local property owners to attempt to stabilize the shoreline by costly structures and in turn such sea-wall construction may intensify erosion further along the shore.

The review of the coastal sector between Seaton and Downderry demonstrates that active marine erosion is ongoing, primarily generated by erosional wave conditions although aided by changes in nearshore sediment supply and the nature of the cliff materials. Despite local (private) attempts at sea-wall construction to protect individual properties, erosion continues due to the piecemeal nature of such undertakings. Owing also to the discontinuous nature of these defences, wave attack works around the side and eventually behind these structures, which ultimately collapse. In some cases the lack of adequate drainage through the concrete or masonry walling has led to a build up of water pressure on the landward side leading to major fractures and failure of the structure. Quite often also, inadequate toe protection provided by deeper foundations or vertical piling leads to these simple sea walls 'toppling over' when even moderate wave action has removed or shifted the narrow zone of beach sediments which offer natural protection to these low-cost schemes (Coard, Sims and Ternan, 1987).

This poses a problem as it is difficult to persuade coastal property owners, users and managers that future development should be restricted if the reason for the restriction is not plainly evident at the time. As a consequence, traditional responses to coastal erosion have resulted from the desire to do something tangible in response to perceived erosion events, rather than from a considered forecast or plan, involving adjustments to both socio-economic as well as natural systems (Ricketts, 1986). To stop erosion completely would require either the continuation of the County Council's sea wall, rock armour, or a stone-filled gabion basket system along the whole length of the coast between Seaton and Downderry. At an estimated cost of over £3 m for a concrete wall and £400 000 for gabions, the funding of such structures is obviously of prime importance. In cases such as this, finance is sought from central financial loans for the provision of remedial works under the terms of the Coast Protection Act, 1949.

This Act empowers maritime District Councils, known as Coast Protection Authorities (CPAs) to construct works to protect land against erosion and encroachment by the sea. Neither of the terms erosion nor encroachment is defined in the Act, but erosion is taken to mean the loss of land while encroachment concerns the subsequent incursion of the sea. With grants awarded to CPAs under the heading of coast protection, County Councils are required to make a generally unspecified contribution (usually 50 per cent) to the net grant cost (HMSO, 1985). In considering particular applications for grant aid, the Government has to be satisfied that schemes are cost-beneficial. The appraisal covers both quantifiable and non-quantifiable aspects of costs and benefits, including conservation aspects, and the schemes have to be cost effective overall.

In the case of coast protection between Seaton and Downderry, the County Council (in its capacity as Highway Authority) undertook the cost of sea-wall construction to protect likely damage to the coast road. It was envisaged that this structure would be extended under a grant made available to the District Council. As is shown in the section above, considerable debate and subsequent delays have ensued relating to the proposed extension of the sea wall. The basic problem has been and continues to be the cost of the scheme in relation to the value of the properties to be protected.

## Legal Aspects

Existing law in England and Wales relating to the coastal zone is complex with differing Acts relating to a variety of activities, summarized in Table 13.2, and discussed more fully in Chapter 11 by Carr.

An important result of the 1949 Coast Protection Act was that a distinction between coastal protection and sea defence became entrenched within national coastal legislation, with sea defence (tidal inundation) being specifically excluded. As a result, two separate governmental ministries were involved, with coast protection being the responsibility of the Department of Environment while Regional Water Authorities acting under the Ministry of Agriculture, Fisheries and Food, looked after sea defence. Such a jurisdictional and managerial division creates confusion within the decision-making and funding procedures, a situation which the Government attempted to rectify in 1985 by integrating coastal protection and its grant awarding machinery with the scheme operated by MAFF for land drainage and sea defence.

**Table 13.2** Legislation applicable to the coastal zone

| | |
|---|---|
| Coast Protection Act, 1949 | Prevention of erosion and encroachment by the sea |
| Land Drainage Act, 1976 | Navigation, sea defences to prevent flooding |
| Crown Lands Act, 1866 | Mining, quarrying and dredging |
| Wildlife and Countryside Act, 1981 | Conservation of marine flora and fauna |
| Sea Fisheries Regulation Act, 1966 | Control of fishing industry |
| Control of Pollution Act, 1974 | Restriction of discharge into coastal waters |
| Dumping at Sea Act, 1974 | Prohibition of dumping in territorial waters |
| Prevention of Oil Pollution Act, 1971 | Prohibition of discharge of oil in territorial waters |
| Customs act, 1952 | Prevention of smuggling |
| Highways Act, 1959 | Quarrying and navigation |

*Source:* I. H. Townend (1986), Coastal studies to establish suitable coastal management procedures, *J. Shoreline Management*, **2**, 131–54. Reproduced by permission of Elsevier Applied Science Publishers Ltd.

This shift in the task of overseeing coast protection has been rationalized on the grounds of superior project appraisal methods operated by MAFF as indicated by Penning-Rowsell, Thompson and Parker, (Chapter 12). As has been indicated above, this relatively rapid change left Caradon District Council in a difficult position in relation to plans for coastal protection at Seaton/Downderry as MAFF engineers indicated that grant aid would be unlikely for a high-cost scheme. Thus for CPAs, the change in responsibilities at central government level has exacerbated the complexity revolving around coastal protection works. Previously the grant aiding for protection came from the same government department as was concerned with both land planning and environmental conservation. Now a different ministry is involved, necessitating different lines of communication and differing policies (Penning-Rowsell and Thompson, 1987). Another aspect is that individual CPAs, concerned with the maintenance of their own independent authority, have not set up Coast Protection Boards under the terms of the 1949 Act. This regional approach was envisaged to include representatives from other interested parties such as sea defence and conservation. Thus, in terms of erosion management, the coastal system remains divided by local political boundaries, with each District Council responsible for its own stretch of coast and with no mechanism for overall concern or responsibility. However, as a result of the shift in responsibility from DoE to MAFF there should now be a greater conformity in the implementation of coastal protection schemes. Perhaps more importantly it should now be possible to consider planning and management of protective works within regional regimes or domains, rather than a piecemeal consideration of engineering works limited to the jurisdiction of a given administrative authority.

Moreover, the scope of research within the financial aspects of coast erosion is often geared towards supporting proposed engineering works. Proper scientific research into geological controls or geomorphological processes is invariably not carried out by local Borough or District engineering departments or even by consulting engineers. Quite often also, small CPAs do not have the financial resources to pay for such scientific work and thus advice may be sought, in the absence of their own engineering department or other technical expertise from engineering consultants whose advice may be less than independent and may well be biased towards large and costly schemes. It is interesting to note here, that at Seaton–Downderry, the local Parish Council and Residents Association have repeatedly suggested lower cost coastal protection schemes as a possible alternative to a concrete sea wall. It may well be, especially given the current availability of prior discussion under the revised arrangements, that the District Council and MAFF may be able to come up with a scheme which could satisfy local residents and property owners. Certainly, effective process–response modelling is required in order to evaluate real benefits against costs. Up to now, the decision to choose an engineering structure to combat coastal erosion has been mainly political and financial rather than a scientific decision based upon appropriate criteria. In the situation at Downderry, of prime importance should be the realization that coastal erosion is complex. We are not dealing with a simple case of storm wave activity aided by sea-level rise. Sediment supplies from alongshore need to be considered as does the role of groundwater in the head deposits of the terrace. Where such water is augmented by storm drainage soakaway, cliff failures are likely to occur. Planning regulations do not appear to allow an adequate consideration of the possible impacts of proposed development on natural processes such as landslides, through for example the disposal of storm water. A broader perspective is essential in cases such as Seaton and Downderry to enable the adoption of a more integrated approach to the problems of coastal development, erosion and protection.

In this context, McGown and Woodrow (1987) argue that coastal studies must consider lengths of coast which behave as a single unit, including both stable and unstable sections. Such lengths of coastline are termed by them, a Coastal Process Unit, where interactions between the offshore zone, intertidal zone and beach, local sea defences, coastal slopes and the hinterland must be taken into account. Because there is a mutual dependency between the various component parts of such a system, investigations of coastal sites should cover each component part of the Coastal Process Unit in order to identify the system dependency.

A similar approach has been undertaken in the sand budget studies of the East Anglian coast (Clayton, McCave and Vincent, 1983). Based on field research and computer modelling, an outline sand budget was established. This showed that the sandy beaches of Norfolk and Suffolk depend on the erosion of cliffs for their sand supply and thus coastal defence works should aim at maintaining high beach levels. The basic premise was that the best form of coastal protection is a good beach and that at some locations sand feeds are likely to be more economical than improved walls or revetments.

As McGown and Woodrow (1987) have stressed, attempts should be made to maintain a 'status quo' within the Coastal Process Unit. However, success ultimately must depend on two things. Firstly, good background data such as can be provided by both short- and longer-term geomorphological investigations and secondly, a suitable management strategy which builds upon a full geomorphological appraisal. To help achieve effective management, key information concerning coastal processes should be stored in an adequately designed geomorphological database (Figure 13.4).

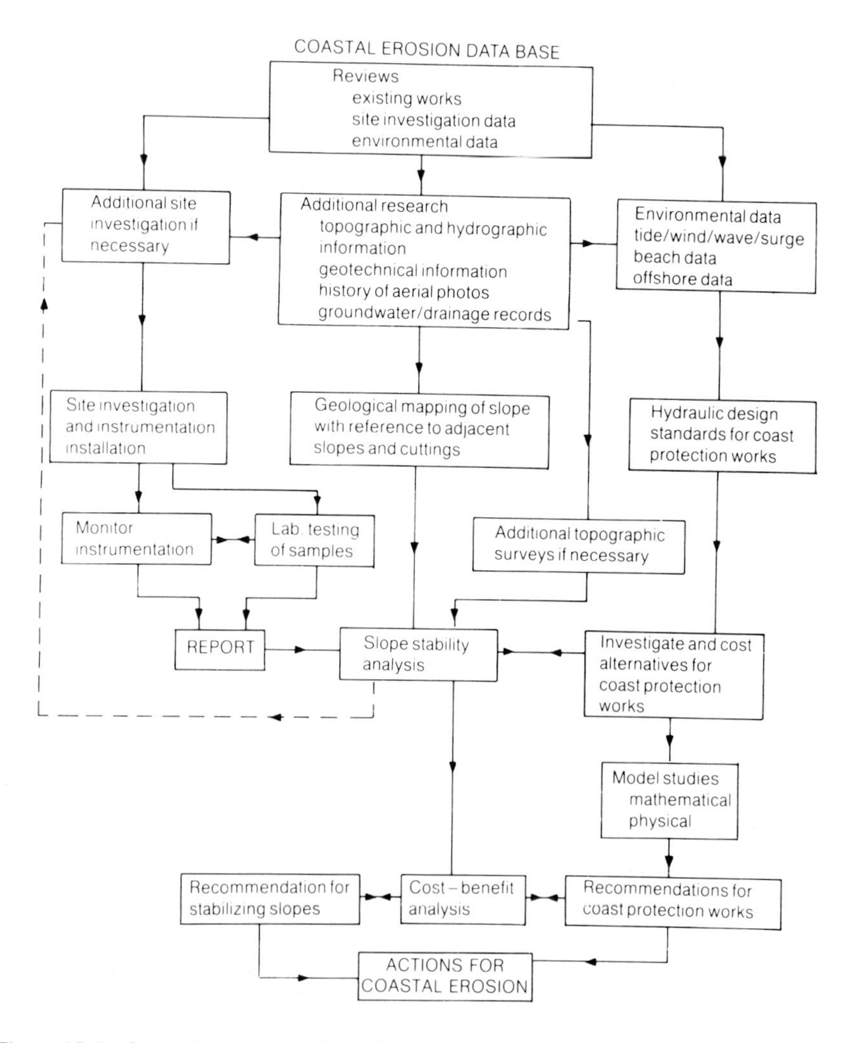


**Figure 13.4** Coastal erosion data base. Taken from Roberts and McGown (1987): A coastal area management system as developed for Seasalter–Reculver, N. Kent. In *Proc. Instn. Civ. Engrs.* Part 1, **82**, 787. Reproduced by permission of the Institution of Civil Engineers

### Planning Aspects

Within the scope of British legislation, land-use management is incorporated within the various Town and Country Planning Acts—under which coastal matters are generally ignored. Planning continues to be thought of in terms of land and sea; town and country (Joliffe and Patman, 1987). Quite often this thinking leads to the total divorce of land management from coastal dynamics and geomorphic processes—despite the fact that engineering structures devised under the terms of the Coastal Protection Act require planning permission. In fact until the planning process gives adequate consideration to the coast beyond the view that the shoreline is simply where land planning ends, this separation of effective management of development and erosion control will remain.

As has been shown at Downderry, planning permissions have been granted by the District Council despite the 'in house' knowledge relating to ongoing cliff recession and problems relating to coastal protection funding. In part, planning departments have their hands tied firstly by legislation which indicates that permissions for development may be granted provided that there is no likelihood of public harm. In this way it is theoretically possible to construct a house of straw in an isolated position on a quaking bog, subsequent damage or loss of the property being the responsibility of the owner. Secondly, planning officers are under constant pressure to process applications quickly. Quite often this leads, given current staffing levels, to perhaps hasty decisions and an inability to investigate further or even take expert independent advice, particularly in matters relating to possible environmental hazards.

## CONCLUSION

The nature of the coast in an area of south-east Cornwall has been detailed together with a review of local erosion and protection problems. A cause for some concern here, between Seaton and Downderry is the apparent lack of a full comprehension of the coastal regime coupled with an overall integrated approach in which nearshore sediment dynamics and other geomorphic processes are evaluated in conjunction with known rates of erosion. These in turn must be related to proposed protection works, storm soakaway drainage and further property developments. The role of the water authority in providing for adequate storm drainage may also have a crucial part to play in the case study cited here.

Geomorphologists with their background training in interactions between physical processes and human impacts over varying timescales, seem to be the most appropriate scientists to advise on policy decisions. Unfortunately, the use of such expertise is quite often a rarity.

Finally, of undoubted use to both local and central government departments would be the development of a national database itemizing environmental hazards. Quite often planners do not have adequate information to hand when making decisions on development, a situation which can lead to a total property loss on the one hand and certainly contradictory decisions and ineffective management on the other.

## REFERENCES


Carson, M. A. and M. J. Kirby (1972). *Hillslope Form and Process*, Cambridge University Press. 475pp.

Clark, M. J., P. J. Ricketts and R. J. Small (1976). Barton does not rule the waves, *Geographical Magazine*, **48**, 580–88.

Clayton, K. M., I. N. McCave and C. E. Vincent (1983). The establishment of a sand budget for the East Anglian coast and its implications for coastal stability. *Shoreline Protection Conference*, Southampton, Thomas Telford, London. pp. 91–6.

Coard, M. A., P. C. Sims and J. L. Ternan (1987). Coastal erosion and slope instability at Downderry, south-east Cornwall—an outline of the problem and its implications for planning. In M. G. Culshaw, F. G. Bell, J. C. Cripps and M. O'Hara (eds) *Planning and Engineering Geology*, Special Publ. No. 4, Geological Society, Engineering Geology Group, 529–32.

Culmann, C. (1986). *Graphische Statik*, Zurich.

HMSO (1985). Financing and administration of land drainage, flood prevention and coast protection in England and Wales, *Cmnd 9449*, HMSO, London. 25pp.

Joliffe, I. P. and C. P. Patman (1987). Problems in the supply of beach resources. In *Coping with the Coast, Proceedings 4th Annual Conference of the Coastal Society*, pp. 161–77.

Lemon and Blizard (1974). *Coast Protection, Seaton-Downderry*. Consulting Engineers Report for Caradon District Council, Lemon and Blizard, Plymouth. 7pp.

Lemon and Blizard (1981). *Report on Seaton–Downderry Cost Protection*. Report for Caradon District Council, Lemon and Blizard, Consulting Engineers, Plymouth. 11pp.

Lemon and Blizard (1984). *Seaton–Downderry Coast Protection Scheme Stage 1*, Report for Caradon District Council, Lemon and Blizard, Plymouth, 7pp.

McGown, A. and L. Woodrow (1987). Planning and geotechnical aspects of coastal landslides. *Conference of River and Coastal Engineers*, MAFF. 18pp.

Penning-Rowsell, E. C. and P. M. Thompson (1987). Coastal erosion and flood control: changing institutions, policies and research needs. *Paper presented at IBG Annual Conference, Portsmouth, 1987*. 24pp.

Ricketts, P. J. (1986). National policy and management responses to the hazard of coastal erosion in Britain and the United States, *Applied Geography*, **6**, 197–221.

Roberts, A. G. and A. McGown (1987). A coastal area management system as developed for Seasalter-Reculver, North Kent. *Proceedings Institution Civil Engineers Part 1*, **82**, 777–97.

Sims, P. C. and J. L. Ternan (1987). *An Initial Investigation into Alternative Lower Cost Coastal Protection Schemes for Downderry and Seaton, Cornwall*, Report for Caradon District Council, Poly Enterprises Plymouth Ltd., 17pp.

Thomas, C. (1985). *Exploration of a Drowned Landscape: Archaeology and History of the Isles of Scilly*, Batsford, London. 320pp.

Townend, I. H. (1986). Coastal studies to establish suitable coastal management procedures, *Journal of Shoreline Management*, **2**, 131–54.

# Policy Formulation

Geomorphology in Environmental Planning
Edited by J. M. Hooke
Published by John Wiley & Sons Ltd



# 14 Geomorphological Information Needed for Environmental Policy Formulation

**DAVID BROOK and BRIAN MARKER**
*Land Stability Branch, Minerals Division, Department of the Environment, London*

## INTRODUCTION

Geomorphology has been defined as the study of landforms. It is concerned with the nature, origin and evolution of landforms, their processes of formation and their material composition. It is predominantly a field science, requiring knowledge of the techniques available for studying landforms, terrestrial processes and surface materials. It demands a broad knowledge of related earth and life sciences, particularly geology, climatology, meteorology, ecology, pedology and hydrology. All these disciplines relate directly to the study of the environment in its widest sense and thus provide information which forms the basis for policy formulation.

Geomorphological information has probably not been applied to the needs of environmental policy formulation to the extent that it could, or should, have been. There is, however, an increasing recognition by both geomorphologists and by policy makers that geomorphology has an important role to play alongside other disciplines in the environmental sciences. In order for that role to be effectively fulfilled, the policy-maker needs to be aware of the information geomorphology can provide and of how it can assist in policy formulation. The geomorphologist too must be aware of who are the customers for the information and of their information requirements.

### Environmental Policy Makers

Environmental policies are made and implemented at all levels in society. All those concerned with development and the use of land and with conservation are contributing either directly or indirectly to policy formulation and implementation.

Private individuals contribute directly by their own actions and indirectly by their consumer and investment decisions, by membership of special interest groups such as trade unions, professional and learned institutions or amenity societies and as electors. Commercial and industrial organizations contribute through their own actions and decisions and through their trade associations. Both private sector and nationalized industries can thus have a major impact on environmental policy at the local, national and even the international level.

More commonly recognized as policy-makers are local parish, district and county councils. In their activities as landowners and developers, as well as in such fields as planning, environmental health, industrial promotion, recreation and land reclamation, local authorities probably have most impact on local environmental policy. Through the local authority associations, they also contribute significantly to regional and national policy formulation.

Non-governmental organizations are major contributors both in formulating their own policies and in influencing the policies and actions of others. They may be government sponsored, such as the Countryside Commission, the Nature Conservancy Council and the Royal Commission on Environmental Pollution or they may be independent of government such as the special interest groups already mentioned.

Within central government, the principal responsibility for environmental policy formulation lies with the Department of the Environment. It is not alone in that responsibility, however. All government departments are constrained by the National Parks and Access to the Countryside Act 1949 to take account of the effects on the environment of their actions, policies and decisions. The Ministry of Agriculture, Fisheries and Food, with its interests in agricultural land capability, land drainage, soil erosion, coast protection and sea defence is also a significant policy maker with an obvious requirement for geomorphological information.

As the principal environmental policy-maker in Government, the Department of the Environment has a wide range of interests. Minerals, water resources, waste disposal, land reclamation, environmental pollution and land stability all fall within the Department's remit. Of special importance is the Department's responsibility for the administration of the town and country planning system, which is concerned with the development and use of land. Since most development takes place on the surface or in the immediate sub-surface, planning policies are concerned with the interface which is the traditional speciality of geomorphologists.

Thus, there are many interests both within and outside government which are directly or indirectly involved in the formulation and implementation of environmental policy. All these different interests may be customers for geomorphological information. Whilst some will not explicitly appreciate that need, many do; some even employ geomorphologists or other earth scientists.

This chapter cannot attempt to cover all the various interests, it can merely point to their existence. It concentrates, therefore, on aspects of interest to the Department of the Environment, particularly in relation to land use planning.

The main emphasis is on issues dealt with in the division responsible for minerals planning and land stability.

## LAND USE PLANNING

The British planning system has not changed in its essentials since its establishment in 1947. The legislation is now contained in the Town and Country Planning Act 1971, as amended by subsequent Acts. The system is designed to regulate the development and use of land in the public interest. It operates to facilitate much needed development, and to strike the right balance between that development and the interests of conservation, in order to secure economy, efficiency and amenity in the development and use of land.

The fundamental requirement of the legislation is that development may not be undertaken without planning permission. Development is defined in the 1971 Act (Section 22) as 'The carrying out of building, engineering, mining or other operations in, on, over or under land, or the making of any material change in the use of any buildings or other land'. Certain works and uses do not constitute development under the Act, the most important to geomorphologists being 'the use of land for the purpose of agriculture or forestry'.

Forward planning is accomplished through Development Plans, with strategic policies expressed in Structure Plans and more specific, Ordnance Survey map-based, proposals in Local Plans. The provisions in such plans are not prescriptive but they provide a basis for decisions on planning applications. They provide a broad and flexible framework to cover the medium term (ten years plus) needs of the plan area within the context imparted by Government policy statements.

The development control system operates within this framework to determine individual applications for permission to develop and use land. It is of its nature more directive and responds to shorter-term issues, though the effects of decisions may be long-lasting. There is always a presumption in favour of the application unless the development would cause demonstrable harm to interests of acknowledged importance. Each case is decided on its planning merits having regard to the provisions of the Development Plan and any other material considerations (Section 29(1) of the 1971 Act). The system is concerned with land-use planning matters, i.e. those relating directly to the physical development and use of land and not to other matters which are the subject of other legislation, though 'in principle . . . any consideration which relates to the use and development of land is capable of being a planning consideration' (COOKE J in *Stringer* v. *Ministry of Housing and Local Government*, 1971).

## GEOMORPHOLOGY AND PLANNING

Planning is concerned with the physical development and use of land. It is thus clear that geomorphological and other earth science information can be an important element both in the formulation of planning policies and in the determination of individual applications.

In order to formulate and justify their policies as expressed in Development Plans, local authorities are required to carry out a survey of their area examining matters which may be expected to affect development of that area and to keep such matters under review. This review should include, amongst other things, 'the principal physical and economic characteristics of the area' (MHLG and WO, 1970).

The contribution that geomorphology can make to such a survey should be self-evident. The physical characteristics of an area are largely dependent on its material composition and on the processes affecting those materials both in the present day and during their evolution. However, the use of geomorphological information has often been implicit in the consideration given to such matters as topography and landscape conservation rather than being an explicit consideration. It has tended to be descriptive only, as an aesthetic appreciation of landscape. It seems, therefore, that due consideration may not have been given in all cases to the processes affecting that landscape and the impact those processes could have on development or the ways in which development might affect geomorphological processes. The relatively common occurrence, for example, of landslides affecting development or being initiated by inappropriate engineering suggests that hillslope processes affecting land stability are not considered in some planning and development decisions. A broad appreciation of planning in Britain suggests to the authors that, apart from minerals and waste disposal, many land use planning policies and decisions do not give due weight to earth science factors alongside other planning considerations such as amenity, conservation, agriculture and social and economic factors. In particular, very few Development Plans cover matters relevant to land stability.

There are two possible reasons for this apparent lack of consideration of important earth science factors. In the first place, town planning, not unnaturally, is concerned largely with the areas where people live and work, since that is where most development takes place. Development in the countryside is generally resisted for reasons of landscape and agricultural conservation. In addition, much rural development related to agriculture and forestry is effectively outside the planning system. Also, the age structure of existing urban buildings and structures and the decline of economic activity in this context has inevitably led to a concentration on economic and sociological aspects of the deprived inner cities. The recycling of urban land to revitalize inner city areas is an important element in central and local government policy.

However, the importance of geomorphology seems much less obvious in the inner city than it is in less urbanized areas, although it can be relevant even there.

The second, and perhaps more significant, reason for earth science factors not receiving their due consideration may lie in a gradual change in the nature and education of the professional planner. Many early planners qualified initially in other disciplines, principally geography, architecture and surveying. Only subsequently did they obtain a professional planning qualification, often after some years of operating as planners. The modern generation of planners, however, are increasingly qualified solely in town and country planning without the initial qualification in other disciplines. In addition, the professional qualification of the Royal Town Planning Institute formerly included a compulsory section on geology; this is no longer a compulsory requirement. Such an element in the formal training of planners at least gave them an appreciation of the possible impacts of earth science factors in planning. Regrettably such an appreciation may now be largely lacking from the training of a significant proportion of planners.

It is thus important that, in order to maximize the use and consideration of relevant geomorphological and other earth science information, there must be communication between the providers of that information and the potential users. Over the past few years in particular, this has been an important element in the Department of the Environment's Planning Research Programme.

## APPLIED EARTH SCIENCE MAPS

Conventional geological, geomorphological and pedological maps are readily usable to provide information necessary to efficient planning. They are, however, specialist maps which require specific expertise for their interpretation. In consequence they may be almost unintelligible to the non-specialist and may thus not provide an adequate basis for the consideration of earth science factors by planners.

Recognition of this situation by the Department of the Environment and the Scottish Development Department, together with officers from the then Institute of Geological Sciences, led the Department to commission an environmental geology mapping exercise in Glenrothes New Town (Nickless, 1982). This initial exercise was based on an analysis of existing data to separate out in mapped form the individual elements of the geology and to recombine them in a form that was more directly usable by planners.

Subsequently a number of further studies have been commissioned to investigate the best means of collecting, collating and presenting earth science data of direct applicability in planning. The objective has been to present such data in a form which can be readily appreciated by planners to increase the possibilities of due account being taken of the physical opportunities and constraints of an area alongside other planning considerations. The main emphasis has been placed on maps since these are a common form of communication amongst both planners and earth scientists and are readily applied in area studies.

Commissioned studies have produced results which have usually comprised:

(a) A set of maps depicting collected factual information on individual elements of the geology—e.g. distribution of rocktypes, thicknesses of units of special interest, mined ground and mineshafts, landslides, made ground, contaminated land, geotechnical properties.
(b) Interpretative maps drawing on the basic data to define characteristics of particular interest—e.g. extent of possible mineral and water resources, liability to subsidence or landslipping, foundation conditions.
(c) A further step, required in some cases, may be the production of maps compiled from these basic and interpretative sets which summarize the general characteristics of the area in terms of resources for and constraints on development.
(d) An area database of earth science information relevant to planning and development, which acts as a source of primary information and is readily updatable.

The aim in funding this research has not been to attempt to cover the whole, or even substantial portions, of the country with such maps but to develop the techniques of assessment and presentation of earth science information in varied terrains with different geological and planning problems. Such techniques might then be used by those with a direct interest in specific areas, such as local authorities, where they consider it appropriate to do so. Given the relevant earth science information in a readily usable form there is much more chance that it will be used, not necessarily to override other factors in planning, but to be considered alongside such factors and given a weighting appropriate to the particular circumstances (Brook and Marker, 1987).

These applied earth science maps have developed with time. Early work such as Glenrothes, was based on existing data and tended to be strictly geological in its content. It did not produce direct information on the dynamics of the landscape and was thus weakest in presentation of the earth surface features and processes which are the interest of the geomorphologist. Later work has included primary mapping of geology and of other features such as ground contamination, landslides, soil erosion, etc. The importance of the geomorphological contribution to applied earth science mapping will undoubtedly continue to develop; significant steps forward may be a current study by Geomorphological Services Ltd in the Torbay area and work by the Soil Survey of Great Britain in the Southampton area, which is intended to complement that already carried out by the British Geological Survey.

## GEOMORPHOLOGY AND POLICY

There are a number of specific policy issues to which geomorphological information can make a direct and significant contribution. Amongst those of particular interest to the Department of the Environment's Minerals Division are considerations relating to mineral resources, and land stability.

### Mineral Resources

The general aim of minerals planning in Britain is to secure the availability of adequate supplies of mineral resources to meet industrial needs with minimum disturbance to the environment. This requires a balance to be struck between the need for minerals extraction and the effects of extraction on the environment. A balance is also needed between the demands for other development and the avoidance of sterilization of mineral resources.

Problems are most acute for supplies of bulk minerals, notably concrete aggregates, which are produced in large quantities and the extraction of which affects considerable areas of land. These problems are not helped by the generally close correspondence between resources of sands and gravel and high quality agricultural land, and between hard rock resources and upland areas of conservation and amenity value.

DoE Circular 21/82 sets out general supply policies, with indicative figures of expected levels of supply by region. Within this strategic context, which has been developed with the co-operation of the minerals industry and local authorities, applications are made by the industry for areas selected on commercial grounds, taking account of local authority planning policies. Such policies might include the definition of areas of search where mineral extraction would normally receive favourable consideration, of mineral consultation areas where the unnecessary sterilization of mineral resources is to be avoided, and of conservation areas where mineral extraction would normally be opposed.

However, the full range of policy options can only be considered if adequate information on the extent and quality of mineral resources is widely available. To meet this need, the Department has commissioned over 150 mineral assessment studies since the late 1960s covering a large proportion of the major aggregate resources of England and Wales. Most of these studies have been strongly biased towards the geology of the resources with assessment boreholes sited on a grid pattern within the resource area under consideration.

A recent study of glacial and fluvioglacial sands and gravels to the south of Shrewsbury (University of Liverpool, 1985), has made beneficial use of geomorphology in designing the assessment programme and a similar approach is now being applied to parts of northern Gwynedd. The value of geomorphological mapping in targeting assessment to those landforms most likely to contain resources and in assessing the probable consistency of composition of the deposits in this area (and in the Soar and Wreake valleys in Leicestershire; Engineering Geology Ltd, 1985a–c) has successfully been demonstrated.

Whilst the emphasis of the Department's mineral assessment programme is now moving away from sand and gravel towards other minerals, there will remain a need for the significant contribution of geomorphological information to the study of Quaternary and recent deposits in the remaining areas where sand and gravel assessment might be required.

Minerals planning relates not only to the identification of mineral resources but also to the environmental effects of extraction. Here too the geomorphologist has a part to play. For example, the work by Gagen and Gunn (Chapter 8) on controlled blasting of limestone quarry faces to simulate natural landforms shows great promise for the achievement of rapid restoration of at least a proportion of hardrock quarries which may otherwise remain as prominent visual scars on the landscape. Further research on this topic is in preparation. It could well be that the geomorphologist can contribute in a similar manner elsewhere to the speeding up of the natural processes of regeneration which enable quarries to be assimilated in the landscape.

A recent study by the Geoffrey Walton Practice (1988a) has reviewed the stability and hydrogeology of deep mineral excavations. Again geomorphology has made a significant contribution alongside geotechnics and hydrogeology to such topics as rockfall and other forms of slope instability, karst phenomena and surface runoff from these artificial catchments. An important element of this study was the development of guidelines on the investigation, assessment and design of quarry faces (Walton and Brook, 1987; Geoffrey Walton Practice, 1988b). These are intended to assist the industry, those with an interest in land near the mineral excavation and local planning authorities in ensuring that adequate consideration is given to matters of stability and hydrogeology.

## Landslides

Landslides can have a significant impact on people, properties, structures and communities in general. Because of this and in pursuance of its administration of the planning system, the Department of the Environment has long recognized its role in reviewing the scale and extent of problems arising from landslides and assessing the general applicability of methods and techniques developed at specific sites. As a result, it has funded a number of studies of landslides in different areas and situations.

It was recognized by the Department and the Welsh Office that landsliding was of particular significance in the valleys of the South Wales Coalfield. Research in that area has followed a three-phase strategy:

(1) Survey of the problem—to estimate the size, seriousness and geographical distribution of problems arising from landslides (Conway *et al.*, 1980);

(2) Studies of causal mechanisms—to identify the factors that have contributed to the occurrence of landslides and to make a preliminary assessment of their relative importance (Conway *et al.*, 1983);
(3) Assessment of landslip potential—to assess the potential or susceptibility of slopes for landsliding for the benefit of land use planners and developers (Sir William Halcrow and Partners, 1986).

With the development of this strategy, geomorphological information has been able to contribute significantly to land use policies so that areas with potential for movement may be avoided or any necessary preventative or remedial action may be taken in advance of construction. Planners and developers are also alerted at the earliest stage of any need for intensive site investigation, of the form that such investigation might take and of the broad cost implications of developing specific areas.

The Landslip Potential Planning Map (Sir William Halcrow and Partners, 1986) is regarded as particularly significant in that it goes beyond the mere identification of existing landslides to identify areas where hazard may arise if the present apparent slope equilibrium conditions are disturbed. More importantly, it not only identifies the potential for slope movement but it also tells planners how to use the map. This is about to be tested in practice by Rhondda Borough Council along with the completion of landslip potential mapping for the Borough and the development of guidelines for planners and developers on the considerations which need to be incorporated in the planning system.

The landslip potential mapping so far developed relates principally to shallow landsliding and, whilst it represents a significant advance in the consideration of landslides in planning, there is no specification of the factors that have triggered either first-time movement of landslides or the renewal of movements of ancient landslides which have long been stable. In particular, the significance of undermining could not be unequivocally assessed, though there was evidence of a temporal and spatial relationship between mining activity and landslide movement elsewhere in the coalfield (Sir William Halcrow and Partners, 1985). Consequently research has recently been commissioned on the relationships between landslides and undermining and further research is to be undertaken on the precise factors that have triggered movement of landslides in the South Wales coalfield.

Elsewhere in Britain, the problems arising from landsliding were not so clearly defined in terms of their significance, the measures necessary to take account of them nor the priority areas where further work might be necessary. The Department thus recognized the need for a comprehensive examination of landsliding in Great Britain to identify the scale and nature of the problems and how to overcome them. Geomorphological Services Ltd (in association with Rendel, Palmer and Tritton) were therefore commissioned to carry out a review of research (and practice) on landsliding in Great Britain.

The broad aims of this review were to determine what work had been done on landslides, to assess its general applicability and to identify gaps in knowledge as well as priority areas where work may be needed to assess any risk to communities, properties and structures. Specific objectives related to the geographical (and geological) distribution of landsliding, causes and mechanisms, investigation and remedial measures, the quantification of landslide hazards, the legislative and administrative provisions relating to landslides and the need for further research.

This review is now nearing completion (Geomorphological Services, Ltd, 1986–87). Of the order of 7000 landslides have been identified from the published literature and a database of published information has been established, keyed into regional atlases and a bibliography of principal references. This database has now been made available for consultation by interested parties.

Perhaps more significant than the amount of data is the generally poor quality of that data, not only in relation to the distribution of landslides but also to the information recorded on the vast majority of individual landslides (with the exception of a limited number of well known and spectacular examples).

Another cause of concern was the consideration given to landslides in the planning system. Apart from a few isolated examples it was found that few planning authorities gave adequate consideration to matters of slope stability in either development plans or development control. A primary need was identified to increase awareness amongst planners, engineers, developers, insurers and landusers in general of the large amount of slope instability in Britain and of the costs that arise when this knowledge is not taken into account.

The Department will be reviewing its policy and research needs in the light of the findings of the review and its recommendations. It has, however, already signalled its intention of funding work on landslides and planning in the Ventnor area of the Isle of Wight undercliff where substantial urban development exists on active coastal landslides.

## Underground Cavities

Underground cavities may cause problems for construction and for mineral extraction. Such cavities may have a natural origin (e.g. solution cavities, tension gulls), or they may be due to mans' activities (e.g. mining, civil engineering, construction). While their origin may be different, the processes leading to surface effects are essentially similar and the surface landforms which result may be virtually identical. Geomorphological mapping and aerial photograph interpretation can therefore assist in providing information needed for policy purposes.

Some areas where underground cavities cause problems are reasonably well known. Others are not and both the cavities and consequent surface landforms may not be readily identifiable prior to problems arising. Whilst such incidents, construction problems and remedial measures are periodically reported in both the technical and the popular press, the scale of the problems and the cost implications seem to be imperfectly known.

To remedy this situation, the Department is about to commission reviews of both mining subsidence and of natural underground cavities. The general aims and objectives of these studies are similar to those of the landslides review.

## Foundation Conditions

The design and costs of foundations and the problems encountered during building and construction depend in large measure on the relationships and properties of near-surface deposits. Over large parts of Britain this involves consideration of superficial deposits which have been extensively studied and interpreted by geomorphologists, though often for purposes other than site investigations. Only recently has engineering geomorphology become a rapidly growing sub-discipline.

Once again the overall scale of problems and their financial implications are not clearly known. However, subsidence claims in the private sector totalled about £90 million in 1986, equivalent to about 10 per cent of the cost of new foundation construction or 0.5 per cent of the annual value of orders for new building and civil engineering construction. Thus problems due to foundation conditions are clearly substantial.

Some aspects of foundation conditions have been considered in commissioned geological studies of specific areas such as Sefton, Plymouth and Bristol, in current work on the Torbay area and in the study of engineering geological aspects of the coastal levels of the Severn estuary which is in preparation. Whilst these are a valuable contribution to knowledge within those specific areas, they do not provide a basis for policy formulation on a national scale. The Department proposes therefore to commission a review of foundation conditions in Great Britain to obtain a clear but general picture of the scale and nature of problems in order to determine the need for and point the way towards the appropriate policy response.

## Water Erosion and Deposition

The importance of water to planning and development relates most obviously to water as a resource, its protection from pollution, and to flooding, which is probably the most costly of the 'geological' hazards in Britain. Less obvious but also of considerable importance are the effects of water in ground instability through hydrostatic support of underground cavities, erosion and deposition.

Problems may be alleviated or compounded by local drainage and protection measures since these may have wider implications than originally intended. Here the geomorphologist's broader views of drainage patterns and processes and their implications may give the wider context for those primarily concerned with more localized matters.

## CONCLUSIONS

This chapter has concentrated on the information requirements for land-use planning, emphasizing geomorphological aspects of interest to the Department of the Environment's Minerals Division. These interests relate primarily to the extraction of minerals and to the consideration of earth science matters in planning, particularly with respect to land stability.

The British planning system is a flexible rather than a directive system, with the forward planning in Development Plans providing a strategic and policy framework for development control decisions on individual applications for planning permission.

Geomorphological information can provide an important element, alongside more conventional planning considerations, in both development planning and development control. In particular, it is increasingly being seen as important in the development of applied geological mapping to identify resources for and constraints on development and in the assessment of mineral resources.

A primary need for geomorphological information has been, and will continue to be in the field of land stability where both ancient and present land surface processes have significant effects on matters such as landsliding, karstic phenomena and foundation conditions. Information is required both to determine the need for and to develop an appropriate policy response. In consequence, a series of reviews—of landsliding, instability due to mining, natural cavities, deep mineral excavations, mine openings and foundation conditions—are in progress to determine the scale, nature and extent of problems arising, and how to investigate and overcome those problems, both technically and administratively. Relevant information is provided by geologists, hydrologists, soil scientists, engineers and others as well as by geomorphologists. To achieve the full potential of studies the different specialists need to work together in multidisciplinary teams and to communicate to the potential users of their information, who will have varying levels of understanding. It is essential that potential users are fully consulted as the results of studies are being prepared.

It must also be emphasized that environmental policy makers, both inside and outside Government have limited time and resources. Thus, whilst political and social sensitivity, costs to Government or the occurrence of significant incidents may raise the policy profile of particular issues, this often means that other issues necessarily take a lower priority. Hence a comprehensive cover geographically or by theme is unlikely, certainly in the short and medium term, and it is necessary to define key objectives for any particular study. Research requirements will often be defined in part on the urgency of the need for information. Similarly interest groups who are pressing for action or research on specific issues should realize that they are competing for finite resources. This emphasizes the need to identify the key issues for study and to relate them to policy needs. The Department of the Environment and other environmental policy-makers (and funding agencies) thus need information on what are the key geomorphological issues.

Finally, the authors wish to emphasize in the strongest possible terms the need for communication. One cannot expect non-specialists to appreciate their own needs for policies and the information on which they are based unless they are educated in the relevance of the work and it is presented in terms which they can understand. Nor can specialists expect to know what information is required by policy makers unless they are aware of how those policies are formulated and implemented. A two-way process of education and communication is therefore required in the needs of policy makers and in appreciation of what is on offer.

### Acknowledgements

The views expressed in this chapter result from many discussions with colleagues in the Department of the Environment and outside, particularly with various research contractors. The views expressed, however, are those of the authors alone and do not necessarily reflect the views of the Department of the Environment.



## REFERENCES

Brook, D. and B. R. Marker (1987). Thematic geological mapping as an essential tool in land-use planning. In M. G. Culshaw (ed.) *Planning and Engineering Geology*, Proc. 22nd Annual Conference Engineering Group, Geological Soc. Land, Plymouth Polytechnic, 8–12 September 1986. Geological Soc. Sec. Publication.

Conway, B. W., A. Forster, K. J. Northmore and W. J. Barclay (1980). South Wales coalfield landslip survey, *Institute Geological Sciences, Special Surveys Division, Engineering Geological Unit, Report*, No EG 80/4.

Conway, B. W., A. Forster, K. J. Northmore and W. J. Barclay (1983). A study of two landslipped areas in the South Wales coalfield, *Institute Geological Sciences, Engineering Geological Unit Report*, No EG 83/6.

Department of the Environment (1982). *Guidelines for Aggregates Provision in England and Wales*, DoE Circular 21/82.

Engineering Geology Ltd (1985a). *Research Programme to Assess the Potentially Workable Sand and Gravel Deposits of the Soar Valley, Leicestershire: Final Report*, Report to Department of Environment.

Engineering Geology Ltd (1985b). *Research Programme to Assess the Potentially Workable Sand and Gravel Resources in the Wreake Valley, Leicestershire*, Report to Department of Environment.

Engineering Geology Ltd (1985c) *Boreholes for Geological Research into the Glacial Evolution of the Wreake Valley, Leicestershire*, Report to Department of Environment.

Geoffrey Walton Practice (1988a). *Technical Review of the Stability and Hydrogeology of Deep Mineral Excavations*, London, HMSO.

Geoffrey Walton Practice (1988b). *Handbook on the Stability and Hydrogeology of Deep Mineral Excavations*, London, HMSO.

Geomorphological Services Ltd (1986–87). *Review of Research into Landsliding in Great Britain*. Reports to Department of Environment.

Ministry of Housing and Local Government and Welsh Office (1970). *Development Plans. A Manual on Form and Content*, London, HMSO.

Nickless, E. F. P. (1982). Environmental geology of the Glenrothes district, Fife: description of 1:25,000 sheet No 20. *Institute of Geological Sciences Report*. No 82/15.

Sir William Halcrow and Partners (1985). *East Pentwyn and Bournville Landslips Research Project*, Report to Department of Environment.

Sir William Halcrow and Partners (1986). *Assessment of Landslip Potential—South Wales*, Report to Department of Environment.

University of Liverpool Department of Geology (1985). *Assessment of Sand and Gravel Resources to the South of Shrewsbury*, Report to Department of Environment.

Walton, G. and D. Brook (1987). Guidelines for designing and inspecting excavated quarry slopes. In M. G. Culshaw (ed.) *Planning and Engineering Geology*, Proc 22nd Annual Conference Engineering Group, Geological Soc. Lond., Plymouth Polytechnic, 8–12 September 1986. Geological Soc. Spec. Publication.

# Conclusion

Geomorphology in Environmental Planning
Edited by J. M. Hooke



# 15 Conclusion: The Way Ahead

**JANET HOOKE**
*Department of Geography, Portsmouth Polytechnic*

The chapters in this volume have demonstrated that geomorphologists have much to contribute in provision of data and in policy formulation and implementation in the environmental planning field. In a number of areas action is very urgent to prevent severe environmental damage or to prevent much greater problems in the future (e.g. soil erosion). In other cases policy is being changed and an opportunity exists for input of sound geomorphological advice; this is particularly the case with change of agricultural land use. As development increases so does pressure on land, so zones subject to hazards such as flooding or instability are used to a greater extent. Planning based on sound knowledge of geomorphological processes and characteristics in an area can even be more cost-effective in the long run. It is undesirable that action should only be taken once a disaster has occurred.

The need for geomorphological involvement in the planning process occurs at various stages. In some cases, as demonstrated in various examples, ignorance of a land situation has led to problems (e.g. housing development of unstable slopes) but this could be overcome by provision of geomorphological maps and by requirements for geomorphological investigations and checks on sites. Secondly, often the consequences of a development or activity, particularly off the site have not been realized or considered. Geomorphologists' training in analysis of systems and understanding of their interconnections as well as knowledge of the effects of specific activities mean they can predict such effects. Thirdly, it is increasingly realized that the most effective policies and measures are often those which work 'with nature' rather than against it as shown particularly in the fluvial and coastal fields. The techniques discussed for soil erosion prevention by geotextiles and for restoration blasting of quarry faces also mimic nature. Effective design and implementation must be based on geomorphological knowledge. Some measures are again proving cost-effective and are certainly more acceptable environmentally and aesthetically.

There is an increasing realization on the part of both planners and geomorphologists of the need for involvement but strategies for successful collaboration still need to be worked out in many cases. There are several examples in this book of successful collaboration and much of the advice is derived from experience. It has been shown that provision of information within the policy context is essential; it is insufficient simply to provide data and hope it will be used, and used in the right way. It must be communicated to the appropriate people so a knowledge of the decision-making process is required by geomorphologists. It must also be communicated in an understandable form. It often means that scientists cannot remain aloof and 'unbiased' but must actually take a policy stand. In the longer term greater environmental education is necessary to ensure awareness on the part of planners, and all involved in the environment, of the need for geomorphology.

The decision-making and policy-formulation processes and the institutions themselves do present some problem, with a lack of coherence in some fields. At present in Britain the role of planning and degree of local control is also being decreased so this may present even greater problems in gaining effective policies in the future. More and more development is taking place by private firms and from individual initiative rather than public policy. Frameworks for inclusion of geomorphological considerations are often inadequate and responsibilities for environmental consequences not borne by the developers. Much can be learnt from experiences in other countries.

One of the reasons for the lack of geomorphology in environmental planning in Britain up to now may be the perception that generally Britain is a relatively stable environment, that most geomorphological processes are not rapid and that the land is not subject to the extreme events and hazards of other environments. Several chapters in this volume have demonstrated that this image of stability may be mistaken especially with the increasing anthropogenic influence on processes and forms. Cases of severe soil erosion, landsliding, rapid cliff recession and eroding river channels have all been cited, many induced or exacerbated by human activities. The problems may be even greater in other countries and there, geomorphologists can make a correspondingly even more significant contribution.

A number of specific conclusions can be drawn from the chapters in this book. Newson (Chapter 2) has shown the need for clear policies on upland land use and for decisions to be made based on knowledge and advice of the environmental effects, much of which is already available. The urgent need for action to reduce and prevent soil erosion in Britain is demonstrated by Boardman (Chapter 3) and he suggests some specific strategies for action. He points out that lack of action may have long-term effects on soil productivity. Morgan and Rickson (Chapter 4) discuss one means of erosion reduction but indicate how development and use of geotextiles must be based on understanding of processes and interactions at the earth's surface.

The immense problems which can arise in urban areas are reviewed by Douglas (Chapter 5) and he suggests that geomorphology should be explicit in local planning controls. Our historical inheritance in cities as well as present activities may lead to high metal levels in soils which can be a health hazard according to Aspinall, Macklin and Openshaw (Chapter 6). Brunsden (Chapter 7) advocates that the problems posed to development by slope instability could be overcome by adequate mapping to provide information on unsuitable areas. He is already involved in provision of such information. Gagen and Gunn (Chapter 8) make specific proposals which could help to reconcile mineral exploitation with conservation and aesthetics.

The consequences of river channelization are discussed by Brookes and Gregory (Chapter 9) and the effectiveness of such engineering schemes is questioned. Strategies for greater use of natural forms and processes are suggested. Crabtree (Chapter 10) also demonstrates the benefits of a more general and environmentally-based approach to river pollution. In coastal areas all three papers (Chapters 11, 12 and 13) have stressed the problems caused by multiple authorities and dispersed responsibilities, by conflicting or non-existent policies and the lack of awareness on the part of planners of the nature and rate of coastal processes. There is a dearth of adequate information and much potential for geomorphological work.

Thus there are three main areas of scope for geomorphologists in environmental planning. Some areas already exist in which there is sound knowledge and a good database. Such information must be communicated effectively to the policy-makers. Secondly, there are subject areas in which the geomorphological methods and expertise are highly developed but which need to be applied to specific geographical areas to provide specific data. Planners should be made aware of the techniques and expertise on offer. Thirdly, there are some areas where there is an information and policy void and fundamental research is needed to help formulate policy. Geomorphologists can aid in both the identification of such areas and the fundamental research.

The environmental consequences of planners and policy-makers remaining ignorant of geomorphological issues may be damaging and even dire. Geomorphologists owe it to society to communicate their findings and use their expertise. The politicians have to be influenced to provide an adequate framework for the input of such information.

# Index

**D**

**Q**



**R**